Forest Biotechnology in India

Editors
S.A. ANSARI
C. NARAYANAN
A.K. MANDAL

2008

SATISH SERIAL PUBLISHING HOUSE
403, Express Tower, Commercial Complex, Azadpur, Delhi - 110033 (INDIA)
Phone : 011-27672852 Fax : 91-11-27672046
E-mail : info@satishserial.com, hkjain1975@yahoo.com
Website : www.satishserial.com

Published by :

SATISH SERIAL PUBLISHING HOUSE
403, Express Tower, Commercial Complex, Azadpur, Delhi - 110033 (INDIA)
Phone : 011-27672852 Fax : 91-11-27672046
E-mail : info@satishserial.com, hkjain1975@yahoo.com

ISBN 81-89304-56-9

Printed at
SALASAR IMAGING SYSTEMS, DELHI-35

Preface

Forest science in India is very old and has been, by and large, colonial consumer oriented, which encouraged extraction of forest resources, particularly timbers, medicinal plants, paper pulp, etc. as well as identification of diseases, insect pests and their management in the field and forest depot. Consequently, it fostered reckless exploitation of our forest resources to the extent of their decimation, which underlines the significance of replenishment and productivity amelioration to the level of sustainable utilization, environmental security and climate stabilization. The realization has led to emergence of forest genetics and population analysis, which converges into recent advances in biotechnology.

In comparison to various forest rich countries of the world, forest biotechnology in India remains a late starter and is not holistic. For example, our efforts remain directed towards cloning of superior genotypes and molecular diversity analysis at individual and/or population level. On the other hand, the international researches on forest biotechnology has advanced to the level of QTL identification, construction of genome maps of selected timber species, crisscross of valuable genes, etc. Besides, micro-array analyses of bio-process/ organogenesis have also taken lead in certain cases.

Tropical forests are repository of valuable gene pool, which needs to be explored thoroughly and can be judiciously utilized for improvement of quality and productivity of existing forests/ plantations. Keeping these points in view, it became pertinent to bring all forest biotechnologists together for deliberations to share their research endeavors and chalk out future course of action. In consonance, Genetics and Plant Propagation Division, Tropical Forest Research Institute, Jabalpur organized a two-day national conference on "Tree Biotechnology: Indian Scenario" on February 2005, which was attended by 75 leading forest biotechnologists of the country. They presented about 50 research papers. Of these, the reviewers recommended 30 research papers to be published as an edited volume for benefits of students, administrative authorities and researchers. The first paper presents research status of forest biotechnology in the country, followed by 12 research papers dealing with molecular diversity analysis of eucalypts, poplar, teak, *Albizia* spp., etc, employing RAPD, ISSR, AFLP and isozyme markers. Some of the papers were way forward to address allelic diversity of individual gene. The subsequent 12 papers dealt with somatic embryogenesis and micro propagation of tropical timbers, bamboos, acacias, Indian drum tree, etc; whereas the remaining five research papers endeavored to analyze disease resistance at genetic and molecular level and positive role of bio-fertilizer for boosting forest growth and productivity.

All these papers were critically reviewed and prepared to cater to the need of wide audience. The editors expect that the information contained herein will serve as a strong reference point to foster research advancement in the field of forest genomics and proteomics and help construct high density genetic maps of economically important timber/ non-timber forest species that will go a long way in alleviation of poverty of tribals, forest villagers; ensuring sustainable supply of raw material to forest based industries; improvement of environment and arresting of climate change.

We express our deep sense of gratitude to Department of Biotechnology, Ministry of Science technology, Union Government of India, New Delhi, Indian Council of Forestry Research and Education, Dehra Dun and Society for Promotion of Tropical Biodiversity, Jabalpur for providing financial support to organize the event and to Shri I. Samson for type setting of manuscripts. We are also indebted to reviewers, who spent their valuable time to review and improve the manuscript. Finally, we take all responsibility for any omissions while editing the volume.

Place: Jabalpur

S.A. ANSARI
C. NARAYANAN
A.K. MANDAL

Contents

Forest Biotechnology in India © 2008
Editors: S.A. Ansari, C. Narayanan & A.K. Mandal, pp.1-24
Satish Serial Publishing House, Delhi

1

Forest Biotechnology– Prospects and Perspectives

K. GURUMURTHI

62-4 Leela Apartments, 2nd Block, Ponnayarajapuram
Coimbatore-641001

ABSTRACT

In India, major focus of biotechnology provides innumerable opportunities to forestry sector which is confronted with many challenges. Molecular markers have been used in the assessment of extant genetic resources that can help arrive at rational conservation strategies. Tissue culture is of value for propagation and long-term conservation of endangered species. Plantation forestry caters to meet the growing wood demands. Tree improvement programs for selected species are in place, but the focus is on yield rather on traits. Biotechnological tools like micro and macropopagation are helping in addressing the growing demand for quality-propagules. Biotechnology can hasten tree improvement programs through deployment of molecular marker strategies for early selection, determining hybrid identities, detecting genetic drift, and identifying contaminants in productive germplasm collections. Domestication strategies have thrown new challenges like increased susceptibility to disease and biotic stresses. Desirable traits are adaptation to stressed sites, reduced lignin, controlled flowering, straight bole, self-pruning, etc. Genetic modification approaches in trees can help in addressing the problems through the introduction of available genes from model crop species. India has a rich traditional knowledge base on the utility of plant species for human and animal health management. Application of biotechnological tools to harness these resources can bring invaluable benefits to humankind in the form of valuable medicines and products of utility.

INTRODUCTION

Forests render invaluable environmental services for humankind in the form of pure air and water, and raw materials for human needs like wood, medicinal plants and other non-timber forest produce. The rich forest resources of India had been indiscriminately exploited. Burgeoning population pressure in the post independence era caused disappearance of large tracts of natural forests, resulting in destruction of habitats and endangering the survival of many life forms. While the demands of the growing population for these forestry resources have considerably increased, the plantation forestry, utilizing clonal selections, has not been fully exploited to meet the growing wood requirements. Focused domestication programmes are required for sustainable yield and for development of hybrids for different requirements. Domestication involves breeding techniques supplemented with biotechnological approaches, biofertilizers, biopesticides and improved cultural methods. Productivity is often inversely correlated to stress and disease susceptibility. Hence, biotechnological tools are required for integrating and pyramiding the genes for desired traits. The need for conservationand characterization of the wild germplasm resources with desirable traits has, therefore, gained importance for future utilization in genetic improvement programmes addressing requirements like faster growth, good tree form, and other new concerns of plantation forestry like tolerance to insect pests and diseases, and abiotic stresses. Tools of biotechnology have been successfully applied in annual crops of agricultural importance. Forestry sector offers novel areas for application of biotechnological tools for the benefit of humankind. Biotechnology has the potential for generation of new cultivars, that cannot be produced by conventional breeding. Appropriate biotechnology tools can reduce the problems due to long generation time in tree improvement detect genetic drift (Kidwell *et al.*, 1994) and identify contaminants in productive germplasm collections (Tripathi *et al.*, 2001).

Production forestry and role of ICFRE

With large areas of forests coming under protection, plantations in wastelands, and in agricultural field (agroforestry) gained importance to meet the deficit in the availability of wood for domestic and industrial uses. Realizing the importance of production forestry, strategic activities for tree improvement are in progress in the Institutes functioning under the Indian Council of Forestry Research and Education (ICFRE; Dehradun, India) in collaboration with State Forest Departments. Many institutes under the ICFRE have developed the germplasm and tree improvement methods for mandate species for the states it interacts with and provides services (Table 1).

Table 1: Details of Institutes under ICFRE and their mandatory species.

Name of the Institute	Mandate Species
Forest Research Institute, Dehra Dun	*Eucalyptus, Shorea robusta, Dalbergia sissoo, Acacia, Populus, Leucaena leucocephala,* tropical pines
Arid Forest Research Institute, Jodhpur	*Eucalyptus, Tectona grandis, Acacia, Azadirachta indica, Ailanthus excelsa, Tecomella undulata*
Rain Forest Research Institute, Jorhat	*Gmelina arborea, Tectona grandis,* Bamboos, *Acacia, Dipterocarpus retusus*
Tropical Forest Research Institute, Jabalpur	*Eucalyptus, Tectona grandis,* Bamboos, *Acacia nilotica, Dalbergia sissoo, Albizia procera, Azadirachta indica, Gmelina arborea*
Institute of Wood Science and Technology, Bangalore	*Santalum album, Pterocarpus santalinus, Eucalyptus, Casuarina, Tectona grandis,* Bamboos, *Dalbergia latifolia*
Institute of Forest Genetics and Tree Breeding, Coimbatore	*Eucalyptus, Casuarina, Tectona grandis, Acacia,* Bamboos, *Azadirachta indica*
Institute of Forest Productivity, Ranchi	*Eucalyptus, Acacia catechu, Adina cordifolia,* tropical pines, *Leucaena leucocephala, Gmelina arborea*
Himalayan Forest Research Institute, Shimla	*Pinus roxburghii, Picea smithiana, Abies pindrow, Dalbergia sissoo, Salix, Populus, Cedrus deodara*

Steps involved in developing the genetic resources are exploration, collection, evaluation, breeding, multiplication and conservation. The approaches followed for exotic species include introduction as well. In case of indigenous species like *Pinus roxburghii* the approach followed is based on location, management and maintenance followed by improvement strategies. The approach has been in terms of development of seed production areas (SPA), clonal seed orchards (CSO), seedling seed orchards (SSO), selection of germplasm for progenitors and clonal accessions of high value. In selected species, hybridising orchards are also being established. Several candidate plus trees have also been identified propagated vegetatively and established as germplasm bank in India (Gurumurthi and Kumar, 1996; Kumar and Gurumurthi, 2000; Varghese *et al.*, 2000; Balasubramanian and Gurumurthi, 2001). Several plus trees from stressed environments have been identified that can form the genetic base for various tree improvement activities (Gurumurthi and Kumar, 1996; Kumar and Gurumurthi. 2000, Balasubramanian and Gurumurthi, 2001). Improved seeds from clonal seed orchards and seedling seed orchards of *Casuarina, Eucalyptus* and *Azadiracia indica* are already made available to user agencies for plantation.

Apart from the efforts of ICFRE, Forest Departments and the corporations in different states and many universities having strong schools of forestry have also selected and established germplasm in collaboration with ICFRE. There are several reports of germplasm collection in bamboo from FRI, TFRI, RFRI, KFRI and State Forest Departments. This population/germplasm are highly valuable and need to be conserved, in addition, accessions added.

Tree based industries like AP Paper Mill, Rajahmundry, SIV Industries, Coimbatore, J.K. Paper Mills, Rayagada, Seshasayee Paper Board, Erode, West Coast Paper, Dandeli, ITC Bhadrachalam, Harihar Polyfibres, Karnataka, Hindustan Newsprint Ltd., Kerala have collected germplasm for their requirements. Amongst the germplasm developed by ITC Bhadrachalam, the clones of eucalypts are widely planted by farmers, corporations and industries.

Role of biotechnology in conservation, evaluation, and utilization of germplasm resources

A tremendous effort has been put in for identification of forest genetic resources. Effective conservation and utilization of the forest genetic resources needs networking, integration and development. Many of these collections can be of use as base population for tree improvement programme and production of high yielding varieties or clones. More importantly, the high yielding clones can form the basic material for development of regeneration protocols and for direct introduction of required genes through transformation approaches. In view of the facts that the trees have long rotation period, need several years for flowering and many are cross pollinated, the clonal material can be effectively used for introgression of genes for different traits. In fact, biotechnology will play much greater role in tree improvement in future. Tree improvement to trait improvement must be the ultimate approach. It can provide trees tailored for cellulose, lignin, sodic and saline soils, even ideotypes for agroforestry. Support is required to inventorise, evaluate and characterise the germplasm and network the research programs for which a massive effort is needed. Converting the germplasm into tree genetic resource is a requirement for production forestry.

The continuous collection, maintenance and addition to the germplasm developed through planting stock improvement is carried out by different Institutes of ICFRE, detailed studies on these germplasms are required to ascertain their broad genetic base. The approach must evaluate and characterise the genetic material to determine utility in terms of genetics, biotechnology, silviculture, resistance to disease and pests.

Thus, investigations are required for:

- Estimating the extent of variability in terms of morphological descriptors and genetic behaviour in relation to environment.
- Determining per cent variability in the germplasm compared to that of natural population.
- Determining the extent of variation that is required to be maintained in exotic germplasm.
- Mass multiplication through macro/micro propagation as well as through development of progenitors of select germplasm.
- Determining germplasm traits for requirement of industry and farmers
- Determining the adaptive traits of species/collection/provenance for salinity, alkalinity, and disease and pest resistance.

The above studies have been seriously constrained by the lack of descriptors. Descriptors based on DNA fingerprinting techniques distinguish individuals for tree improvement programmes, conservation, evaluation, and utilization of the productive germplasm resources.

Studies on distribution of genetic variation for conservation and improvement

To enable long-term survival of the species, conservation strategies should maintain the broadest inter-and intra-specific variations. However, the information on extent of variation in various forestry species is non-existent and consequently the knowledge on what and how much to conserve is lacking. Characterization of extant germplasm resources is, therefore, critical to develop and deploy conservation strategies. DNA markers like Random Amplified Polymorphic DNAs (RAPDs), Inter Simple Sequence Repeats (ISSRs), and Amplified Fragment Length Polymorphisms (AFLPs), which are less susceptible to environmental and developmental influences, have been used for quantifying diversity within and between populations, and for arriving at a core collection of germplasm resources so that criteria of comprehensiveness, adequacy and representativeness are excercised. Molecular markers can be utilized for identifying sites having the highest level of variability and individuals having the most divergent genotypes (Drummond *et al.*, 2000). RAPD and RFLP markers have been used to understand the intra- and inter-population genetic variations in the mangrove species, *Excoecaria agallocha.* It was shown that populations from the West and East Coasts of India grouped into separate clusters (Lakshmi *et al.*, 2000). Microsatellite analysis of the genetic structure in another mangrove species, *Avicennia marina,* revealed that most of the variation was between populations (41-71%), and within individuals in the total population (31-49%). There was little variation among individuals within populations (0-10%). A significant level of inbreeding was also noticed in some populations (Maguire *et al.*, 2000).

The genetic variation in tree improvement is often acceptable if the introductions are from provenances. However, it would be highly rewarding to determine the extent of variation in *ex situ* population and further compare with the distribution of genetic variation in *in situ* population (Francisco-Ortega *et al.*, 1992; Loos, 1994; Muluvi *et al.*, 1999). This provides an approach to determine the extent of variation required for population improvement, the extent of redundancy present in the population, which in turn decides the population sizes. In other words, strategies to maintain optimal variation with minimal population size needs to be developed.

Provenance level variations in *Casuarina equisetifolia* using ISSR and FISSR markers was quantitatively assessed (Yasodha *et al*, 2004). The total genetic diversity (Ht) in the species over 40 loci was high at 0.4280. Mean genetic diversity within provenances (Hs) was 0.272. The total genetic diversity was apportioned within and among provenances. A relatively high degree of population differentiation (G_{ST} of 0.365, or 36.5%) was recorded. The majority of the variation existed within the provenances (73.5%). In studies on *Eucalyptus*, diversity estimates were obtained for 15 provenances. Molecular markers grouped the provenances corresponding to proximity of their location in Australia. Possible origins of the Indian selections could be traced to Katherine provenance. Out crossing between *Eucalyptus camaldulensis* and *Eucalyptus tereticornis* could also be detected. Acquisition of germplasm can be optimized through sampling strategies (Kesseli *et al.*, 1991). There is need to determine the extent of genetic variation for species native to our country, viz., teak, *Pinus roxburghii*, *Dalbergia sissoo* etc. to devise *in situ* conservation strategies.

When core collections are used for introduction of accessions to other places, it is important to develop methods for validation of such collections in early stages. In fact, in tree improvement and tree genetics, such studies of validation are extremely important otherwise it will be many years before the discrepancy is detected. It is important to monitor the propagules derived for core collections, whether it is from Seedling Seed Orchards or Clonal Seed Orchards. Molecular markers like AFLP have been shown to be more accurate in such studies than morphological markers (Akagi, 1997). Amongst the molecular markers, SSRs and AFLPs were found to have high value for information content (Powell *et al*, 1996). Molecular markers can also be developed to determine the functional diversity. Molecular markers have been used for classifying genetic diversity in rice germplasm (Akagi, 1997; Virk *et al.*, 2000), soybean germplasm (Powell *et a.l*, 1996), barley accessions (Russell, 1997) and wild bean (Tohme *et al.*, 1996). Such studies are common in breeding programmes and release of varieties in agriculture. No such studies are conducted in forestry species. It is therefore important to carry out such studies in tree germplasm.

Genotyping assemblage

Management of tree germplasm requires considerable research inputs through development of descriptors for economic and plantation characteristics followed by genetic analysis, which is further strengthened through marker analyses like RAPD's, allozymes, and microsatellites. These methods are effectively followed in conjunction with morphological traits in many agricultural species for characterization. Such studies are almost nonexistent in forestry, but have become essential in view of plant varieties protection act. The major policy approaches for protection of plant varieties and farmer's right bill is mostly concerned with agricultural crops. As clones will be introduced in larger numbers through macro and micro propagation, there will be an absolute need to study and document the fingerprints of important clones. It will become the basic necessity, as evidence for intellectual property rights as well as to provide adequate support for protection of plant varieties, particularly for tree species.

Twelve superior performers of *Casuarina equisetifolia,* planted on a large-scale were selected for morphological and fingerprinting studies. Applicability of RAPD, ISSR, FISSR and AFLP molecular approaches were evaluated for fingerprinting of selected clones of *Casuarina* (Yasodha *et al.,* 2004). It is thus possible to establish a fingerprint reference library of ISSR and FISSR profiles of the clones of *C. equisetifolia.* This information can also be utilized in identification and verification of genotypes in clonal plantations and clonal seed orchards. Twenty-two economically important clones of *E. tereticornis,* available in the germplasm collection at Institute of Forest Genetics and Tree Breeding, Coimbatore, India were used for DNA fingerprinting studies using RAPD. These clones showing superior pulping traits with wide plantation throughout India need to be characterized and documented. A DNA fingerprint database management strategy was developed for 22 eucalyptus clones using Microsoft Access 2000 for corroborating fingerprints of clones maintained in germplasm (Mathish *et al.,* 2001). Unique fingerprints were obtained for eight clones. All the 101 loci screened with the thirteen primers were monomorphic for clones 100 and 138 indicating redundancy in the germplasm collection. Clones 93 and 99 grouped together with a similarity coefficient of 0.956 indicating possible origin from the same parents. Clones 7 and 128 were grouped together with a similarity coefficient of 0.89.

Quantification of genetic drift

The populations of tree species are maintained at different levels namely seed production areas, seedling seed orchards, clonal seed orchards, hybridizing orchards and clonal gardens. The assemblage of population is an indication of different levels of genetic shift in the population except in case of clonal gardens. The probability of genetic drift in segregating population is high due to cross-pollination. Specifications for prevention of pollen contamination are spelt out. Yet, the genetic contamination and drift in the population needs to be monitored and determined (Kidwell *et al.*, 1994).

Genetic drift studies are essential to maintain the productivity levels. The probability of genetic drift in wind-pollinated population is higher compared to that in insect-pollinated population. Quantitative genetic studies provide approaches to determine such drifts (Schittenhelm *et al.*, 1997). However, in case of tree species where generational cycles are much longer, it is necessary to develop strategies to determine the extent of genetic drift with in the population. Many allozymes markers are used in the study in case of vegetable crops (Loaiza Figueroa and Weeden, 1991). They can also be used in conjunction with RAPD/AFLP studies so that the minimal changes occur in genetic diversity particularly in case of introduced populations.

In land race populations, such approaches need to be followed for introduced casuarinas, eucalypts and acacias to determine the effective population size. This is equally important in conservation strategies (Parzies *et al.*, 2000).

Identification of genetic contaminants

The tree improvement strategies, at advanced levels of population like clonal seed orchards, hybridizing orchards are susceptible to genetic contamination. This is because population of desired accessions is only allowed to breed and the probability of pollens carried through wind dispersal can cause contamination. This can often be monitored through sample studies in seeds derived from well-maintained populations (Tanksley and Jones, 1981). Genetic contamination studies are also essential for clonal accession wherein all clones need to be fingerprinted and maintained and any misidentification can be quickly detected wherever suspicion arises (Steiner *et al.*, 1997). In tree species, over a period of time somatic embryogenesis may become a tool for development of improved / genetically modified population. In such approaches methods need to be evolved to study the consistency and stability of somatic embryogenesis in derived population. Such studies have been carried out in hybrid cottonwood (Han *et al.*, 2000).

Clonal propagation through rooted cuttings and micropropagation is widely used for large-scale *Eucalyptus* plantation programmes because of its ability to fix the desirable traits of mature plus trees. However, when a large number of clones are handled, variations and mixings are commonly confronted which may go undetected in the absence of prominent morphological descriptors. Material from germplasm resources are also used in breeding programmes. The economic implications of such inadvertent variations will be serious as considerable time and money is spent before the mistakes are detected. RAPD and AFLP markers can be used to identify mislabeling in clones maintained through tissue culture.

Development of management strategies

Comprehensive tree improvement programmes are often undertaken by silviculturists but various populations and accessions are introduced/ developed on empirical basis. This is because scientific tools are not available to evaluate and characterize the introductions / populations. This explains why tree genetics / tree improvement all over the world is far behind that in agricultural crop improvement. Though quantitative genetics provides many methods, the traits required that need to be validated statistically is often at variance with what is practiced in the field. Very often the tree improvement practices are constrained by inadequacy of land and inadequate availability of accessions. Therefore, even the best management strategies followed in silviculture does not provide the desired results. Two major problems in tree improvement namely, long years to complete breeding generation, limited accuracy of selection methods of superior trees for quantitatively inherited traits of industrial and environmental relevance can be significantly mitigated by direct identification of genotypes at early stages of tree development using a diagnostic system based on molecular markers co- segregating with traits of interest (Balasaravanan *et al.*, 2005). It is thus essential to develop molecular markers to establish improvement strategies and relate the same to conservation strategies for *in situ* populations (Chalmers *et al.*, 1994; Del Rio *et al.*, 1997; Krauss, 2000).

Utilization

Development of molecular markers for functional diversity and for genetic enhancement is of great importance for industrial plantations and wasteland development. Advances in gene mapping technologies can help identify trees with desired traits like fast growth, resistance to disease and temperature. Molecular markers have been used to expedite the screening of germplasm resources for pest resistance and to facilitate the introgression of resistance genes in to susceptible lines (Mayer *et al.*, 1997; King *et al.*, 1999), for diagnostic purposes (Chen *et al.*, 1998) and for predicting the performance of relevant traits. Gene mapping studies can permit isolation of desired genes in the target tree species. Studies to identify genetic markers are essential to facilitate location of desired genes and introduce them with in and even across the species. Development of mapping populations for desired traits like straight bole, self-pruning ability, disease resistance and other desirable traits is an essential prerequisite. There is need to carry out studies in gene function that can lead to manipulation or introduction of desirable traits that are not normally available in breeding populations.

Plantation managers, whether in the forest department or private nurseries, will increasingly depend on trait improved stock in their seedling / clones, and in such cases it is important to determine the parameters for certification of such stock. In fact, there will be an absolute necessity to define the plant quality standards in terms of physical and genetic quality.

All these investigations will provide leads on

- Methods for determining the value of the tree germplasm
- Economic production of genetic resources
- Development of parameters for determining the economic potential for different end use management.
- Determine the physical and genetical quality of seeds and compare the same with less improved and unimproved population of sub-species taxa.
- Planting stock value derived from improved population.
- Development of nursery practices for quality propagules with defined standards and partially defined pedigrees.

- Quantification of propagules that can be produced annually and the extent of the stock that can be provided for plantation activities.
- Developing package of nursery practices and specific nursery infrastructure for such quality propagules.

This approach will provide basis to develop specific identities for

- Registered propagules (Registration ID for source identified seeds)
- Certified propagules (Certified ID for clonal seed orchards).
- Hybrid propagules (Hybrid ID for full sibs).
- Genetic certification (Genetic ID for clones, micro propagated plants)

Providing certification will help in developing the basis for plant quality standards in nurseries. It will facilitate the State Forest Departments, Corporations and industries to develop standards for release of planting materials. Therefore, an apex agency needs to be established with funds and responsibilities for networking, development, evaluation and characterization of germplasm.

Genetic modification for tree improvement

While the traditional improvement programmes combined with clonal approaches have resulted in improved yields, targeted improvements in qualitative traits like disease resistance and reduced lignin are difficult to achieve through the ongoing open pollination based breeding programmes. Lack of identified genotypes for specific traits, long breeding cycles, together with inadequate knowledge of floral biology, difficulties involved in control pollination and inbreeding depression, limit the use of controlled pollination based approaches for breeding desired combination of traits.

Although transformation in tree species has capitalized on basic research in agricultural crops like tobacco and rice, studies on tree species is challenging. Transformation studies in model species poplar and commercially important *Eucalyptus* gave insight for research priorities for other trees. Transformation studies in poplars have been restricted to a few model hybrids, many of which have been selected for ease of transformation (Han *et al.*, 2000). As in agricultural crops, transformation studies have relied on *Agrobacterium* and particle bombardment based approaches. Pollen transformation offers an alternative route in trees recalcitrant to *in vitro* regeneration (Hay *et al.*, 1994; Martinussen *et al.*, 1994). Experimental field trials of various transgenic tree species are already in progress in many countries.

Wood traits

Genetic engineering has been used for modifying lignin content and composition to reduce pollution during pulping. Antisense technology has been used to suppress lignin biosynthesis genes like caffeate O-methyltransferase (COMT) (van Doorsselaere *et al.*, 1995; Tsai *et al.*, 1998) and cinnamyl alcohol dehydrogenase (CAD) (Baucher, *et al.*, 1996). Long-term field trials in UK and France have shown that poplars expressing CAD and COMT antisense transgenes were healthy, had stable transgene expression, and had better pulping characteristics (Pilate *et al.*, 2002).

Hu *et al.* (1999) reported reduced lignin content and increased cellulose accumulation by the down regulation of 4-coumarate: coA ligase. Shani *et al.* (2001) reported that cellulose binding domain gene under control of the cell elongation - specific promoter in poplar led to significant increase in biomass production. Tuominen *et al.* (1995) reported reduction in vessel size, increase in vessel density, and changes in ray development in poplar transgenic for IAA-biosynthetic genes, *iaaM* and *iaaH*.

In India, many industries use *Casuarina and Eucalyptus* for paper and rayon grade pulp. Thus, altering lignin and cellulose profiles, and fiber traits in them may be a commercially viable prospect. This requires identifying homologous lignin and cellulose biosynthesis genes in *Casuarina* and *Eucalyptus*, and understanding their expression and regulation. However, transgenic trees with reduced lignin should undergo rigorous field trials for wind hardiness and tolerance to diseases and other biotic stresses.

Rooting ability

Yibrah *et al.* (1996) observed extensive proliferation of lateral roots in *Pinus contorta* transformed using *Agrobacterium rhizogenes*. Tzifra *et al.* (1998) reported that *rol*-transgenic *Populus tremula* exhibited extensive root formation in a hormone-free medium, increased root surface area, reduced rooting time, high root: shoot fresh weight ratio and high propagation coefficient of shoot-bud regeneration. El-Euch *et al.* (1998) reported that decreased flavonoid content (quercitrin, myricitrin, flavane-3-ols and proantho cyanidins) in stems of antisense chalcone synthase transformed lines of walnut, was associated with enhanced adventitious root formation.

Biotic stress tolerance

Limitations of pesticide usage in tree species and difficulties in selection and breeding for resistant clones make use of transgenic technology, an additional tool to enhance tolerance to biotic stress. Strategies to incorporate insect resistance have mainly been through the use of insecticidal *Bacillus thuringiensis* γ-endotoxin genes. Poplars transgenic for *cryIA(a)* and *cryIIIA* have shown resistance to the pests *Lymantria dispar* and *Chrysomela tremula* (McCown *et al.*, 1991; Cornu *et al.*, 1996). Devising strategies for controlling pests should draw on basic studies on related pests in agricultural crops. Other insect resistance genes used in agricultural crops include protease inhibitor, α-amylase inhibitor, lectins, chitinases, tryptophan decarboxylase, *Vip3A* and cholesterol oxidase (Shah *et al.*, 1995; Brousseau *et al.*, 1999). Strategies for improved tolerance to crop diseases have been based on expression of pathogenesis related proteins (chitinases, β-glucanases, thaumatin like proteins), altering plant receptors for pathogen elicitors, expression of pathogen induced *avr* genes in plants, pathogen derived toxin detoxifying genes and the use of antifungal / antibacterial/ antiviral genes from other sources (Shah *et al.*, 1995; Punja, 2001).

Abiotic stress tolerance

Poplars transgenic for enzymes involved in abiotic stress tolerance like prxA1 (Kajita *et al.*, 1994), superoxide dismutase (Arisi *et al.*, 1998) and glutathione synthetase (Herschbach *et al.*, 1998) have been reported. Improved transformation systems enabling deployment of many genes into a clone is essential for engineering quantitative traits like abiotic stress tolerance. However, recent reports of transfer of up to 13 genes in rice (Chen *et al.*, 1998a,b), enhanced salt tolerance obtained using AtNHX (Na+/H+ antiport) gene in *Arabidopsis* (Apse *et al.*, 1999), and recent approaches on the use of transcription factors like DREB1A that regulate the expression of multiple genes involved in stress tolerance (Kasuga *et al.*, 1999) have spurred renewed interest in engineering for abiotic stress tolerance.

Advantages over conventional approaches

Unlike conventional breeding, genetic transformation allows the introduction of specific genes conferring desirable traits with minimum alteration in the genome. Breaking of species barriers enables the use of genes from other organisms in cases where identified genotypes with the desired trait are not available for breeding.

Problems in use of non-conventional strategies

Base material

Unlike agriculture crops, domestication of tree species is in the early stages and the base material is highly heterozygous. Pure lines and genotypes identified for desired traits are limited. This coupled with the necessity of transforming seedling-derived tissues has made it necessary to identify seed sources tested for their superior genotype. Seeds from the best *E. camaldulensis* seed source selected through provenance tests have been used for transformation (Ho *et al.*, 1998). Similarly, seeds from elite trees of *E. grandis* x *E. urophylla* plantations have been used for transformation (Cid *et al.*, 1999). Traditional breeding programmes in *C. equisetifolia* have an inbuilt mechanism for operational quality seed production through seedling seed orchards and clonal seed orchards. Integration of genetic transformation into conventional tree breeding programmes through the use of identified seed sources, and genotypes from various trials will help in maximizing the benefits of both strategies.

Regeneration protocols

Woody plants are generally recalcitrant to regeneration restricting production of transgenics only to a few genera. As in agricultural crops, transformation studies have relied on *Agrobacterium* and particle bombardment based approaches. Most of the protocols use seedling-derived tissues (Ho *et al.*, 1998; Cid *et al.*, 1999). Asexual transformation methods using *in vitro* shoot cultures established using stems from poplars maintained in green houses have been reported (Han *et al.*, 2000). Ellis *et al.* (2001) observed that the main challenge for transformation in trees is the lack of efficient methods for regeneration of whole plants from single transformed cells of the most desired genotypes. Micropropagation from mature tissues is still a major problem. As most of the regeneration protocols are callus based, possibilities of somaclonal variation pose a serious threat in transforming elite genotypes (Han *et al.*, 2000; Ellis *et al.*, 2001).

Regeneration methods involving minimized use of hormones, and avoiding callus phase can reduce the problem of somaclonal variation (Ellis *et al.*, 2001). Use of DNA fingerprinting in testing the genetic fidelity of micropropagated plants are well documented (Tripathi *et al.*, 2001).

Field studies of transgenic trees

Field trials of transgenic trees are already in progress in over twenty countries. Ecological risks posed by spread of transgenic into the wild, allergic reactions in humans and effects on fauna have been widely debated. Long generation time of tree species makes it difficult to assess the ecological risks. A combination of demographic data from existing non-transgenic populations, simulation modeling of transgenic dispersal and monitoring field releases can be used to guide current risk assessments (James *et al.*, 1998).

Various strategies for gene containment and excision of selectable marker genes have been attempted in genetically modified crops to minimize ecological risks (Daniell, 2002; Hare and Chua, 2002). Pollen being bereft of chloroplasts, transformation of chloroplast has been advocated to prevent escape of transgene. However, chloroplast transformation technology has not been standardised even for several major crop species (Daniell, 2002). Engineering reproductive sterility in tree species has been attempted to prevent transgene escape through pollen and seeds. Mouradov *et al.* (1998) reported the use of barnase for targeted cell ablation of floral tissues in *Pinus radiata.* Engineering for suppression of floral genes has been attempted (Rothman *et al.*, 2000). The response of overexpression of *Arabidopsis* LFY gene varied widely among *Populus* genotypes, showing precocious flowering in only one hybrid and causing production of anthers rather than carpels in a normal female clone. Earlier, Weigel and Nilsson (1995) reported that overexpression of LFY in aspens caused short, bushy plants that flowered within a year. Such studies may have applications in conventional breeding programmes. Kumar (1995) reported that male trees of *C. equisetifolia* produced more biomass than females. Thus, control of flowering may have applications in increasing biomass.

CONCLUSION

Forest sustainability depends on global interdisciplinary cooperation in tree improvement research to meet projected needs while maintaining a healthy ecosystem. It is becoming inevitable to improve the yield and develop methodologies to enhance yield under diverse conditions such as in wastelands, saline alkali areas and flood prone areas. Tools of recombinant DNA technology need to be applied in tree species to incorporate tolerance to drought, salt stress, and environmental pollutants. Clonal forestry is evolving as yield improvement strategy. The resultant problems are in terms of susceptibility to diseases and pests, and hence the need to integrate pest and disease management strategies as well as to apply the biotechnology tools for resistance to pests and diseases. Furthermore, with the intensification of domestication of few species, tree improvement is evolving into trait improvement. For instance, breeding trees for quality cellulose with reduced lignin content has become essential. Therefore, breeding for traits rather than for weight or volume has assumed significance. Similarly biomass strategy is developing into bio product strategy. For instance, to breed for high oil or cineole content in *Eucalyptus* or for increasing azadirachtin content in neem is more important than an increase in the volume or weight of wood. Tree breeders would take several years to achieve the desired product improvement using conventional breeding methods. Hence the need to develop strategies using biotechnology tools to bring together the desired genes. In addition to developing efficient gene delivery systems, identification of promoters conferring xylem specific expression for lignin engineering, stress inducible root specific expression for enhanced salt tolerance, nodule specific expression for enhanced nitrogen fixation, and floral specific expression for flowering control are essential. Towards all these approaches, there is need to assemble the germplasm, characterise and further develop strategies for breeding or methods to bring together genes of economic importance. These approaches would involve registration and certification of propagules, hybrid validation, and genetic certification. Identified characters and certified germplasm is highly important for future tree improvement and breeding programme. Comprehensive germplasm with adequate variation and representative genotypes needs to be defined and put in place. In fact, in tree species what to conserve and how much to conserve, is it self an unresolved issue. Biotechnology, as applied to forest trees, is still a new science that requires answers to the numerous questions it continuously unveils. Application of biotechnology in forestry sector is a challenging task.

References

Akagi, H., Yokozeki, Y., Inagaki, A. and Fujimura, T. (1997) Highly polymorphic microsatellites of rice consist of AT repeats, and a classification of closely related cultivars with these microsatellite loci. *Theor. Appl. Gen.* 94 : 61-67.

Apse, M.P., Aaron, G.S., Snedden, W.A. and Blumwald, E. (1999) Salt tolerance conferred by overexpression of a vacuolar Na^+/H^+ antiport in *Arabidopsis*. *Science* 285 : 1256-1258.

Arisi, A.C.M., Cornic, G., Jouanin, L. and Foyer, C.H. (1998) Overexpression of iron superoxide dismutase in transformed poplar modifies the regulation of photosynthesis at low CO_2 partial pressures or following exposure to the prooxidant herbicide methyl viologen. *Plant Physiol.* 117 : 565-574.

Balasaravanan,T., Chezhian, P., Kamalakannan, T., Ghosh, M., Yasodha, R., Verghese, M. and Gurumurthi, K. (2005) Determination of inter and intra species relationships among six *Eucalyptus* species based on inter simple sequence repeats(ISSR). *Tree Physiol.* 25 : 1295-1302.

Balasubramanian, A and Gurumurthi, K (2001) Genetic divergence in clones of *Casuarina equistifolia* In: *Proceedings of the Workshop cum Peer Review on Casuarina* (ed. Gurumurthi, K.) Institute of Forest Genetics and Tree Breeding, Coimbatore. pp 73 - 84.

Baucher, M.B., Cabbert, G.P., Dorsselaer, J.V., Tollier, M.T., Petit-Conil, M., Cornu, D., Monties, B., Montagu, M.V. and Inze, D., (1996) Red xylem and higher lignin extractability by down-regulation of a cinnamyl alcohol dehydrogenase in poplar. *Plant Physiol.* 112:1479-1490.

Brousseau, R., Masson, L. and Hegedus, D., (1999) Insecticidal transgenic plants: are they irresistible? *Ag. Biotech. Net.* 1: 1-10.

Chalmers, K.J., Newton, A.C., Waugh, R., Wilson, J. and Powell, W. (1994) Evaluation of the extent of genetic variation in mahoganies (Meliaceae) using RAPD markers. *Theor. App. Genet.* 89: 504-508.

Chen, X.M., Line, R.F. and Leung, H. (1998 a) Genome scanning for resistance-gene analogs in rice, barley, and wheat by high-resolution electrophoresis. *Theor. App. Genet.* 97 : 345-355.

Chen, L., Marmey, P., Taylor, N., Brizard, J.P., Espinoza, C., d'Cruz, P., Huet, H., Shiping, Z., de Kochko, A., Beachy, R.N. and Fauquet, C.M., (1998) Integration, expression and inheritance of multiple transgenes in rice plants. *Nat. Biotech.* 16 : 1060-1064.

Cid, L.P.B., Machado, A.C.M.G., Carvalheira, S.B.R.C. and Brasileiro, A.C.M., (1999) Plant regeneration from seedling explants of *Eucalyptus grandis* X *E. urophylla*. *Plant Cell Tissue Organ Cult.* 56 : 17-23.

Cornu, D., Leple, J.C., Bonade Bottino, M., Ross, A., Augustin, S., Delplanque, A., Jouanin, L., Pilate, G., Ahuja, M.R.(ed.), Boerjan, W. (ed.), and Neale, D.B., (1996) Expression of a Proteinase Inhibitor and a Bacillus thuringiensis Delta-Endotoxin in Transgenic Poplars.In: *Somatic Cell Genetics and Molecular Genetics of Trees,* pp. 131-136.

Daniell, H., (2002) Molecular strategies for gene containment in transgenic crops. *Nat. Biotech.* 20 : 581 - 586.

Del Rio, A.H., Bamberg, J.B., Huaman, Z., Salas, A. and Vega, S.E. (1997) Assessing changes in the genetic diversity of potato gene banks. 2. *In situ* vs *ex situ*. *Theor. App. Genet.* 95: 199-204.

Drummond, R. S. M., Keeling, D.J., Richardson, T.E., Gardner, R.C., Wright, S.D. (2000) Genetic analysis and conservation of 31 surviving individuals of a rare New Zealand tree, *Metrosideros bartlettii* (Myrtaceae). *Mol. Ecol.* 9: 1149-1157.

El Euch. C., Jay, A. C., Pastuglia, M., Doumas, P., Charpentier, J.P., Capelli, P. and Jouanin, L. (1998) Expression of antisense chalcone synthase RNA in transgenic hybrid walnut micro cuttings - effect on flavonoid content and rooting ability. *Plant Mol. Biol.* 38: 467-479.

Ellis, D., Meilan, R., Pilate, G. and Skinner, J.S. (2001) Transgenic Trees: Where are we now? *In: Proceedings of the 1st International Symposium on Ecological and Societal Aspects of Transgenic Plantations* (eds. Strauss, S.H. and Bradshaw, H.D.) College of Forestry, Oregon State University, pp. 113-123.

Francisco-Ortega, J., Jackson, M.T., Catty, J.P. and Ford-Lloyd, B.V. (1992) Genetic diversity in the *Chamaecytisus proliferus* complex (Fabaceae: Genisteae) in the Canary Islands in relation to in situ conservation. *Genet. Resour. Crop Evol.* 39: 149-158.

Gurumurthi, K. and Kumar, A, (1996) Establishment of clone bank of *Casuarina equisetifolia* for tree improvement. In: *Recent Casuarina Research and Development: Proceedings of the Third International Casuarina Workshop* (eds. Pinyopusarerek, K. Turnbill, J.W. and Midgley, S.J.) Da Nang, Vietnam, pp.159-161.

Han, K.H., Meilan, R., Ma, C. and Strauss, S.H. (2000) An *Agrobacterium tumefaciens* transformation protocol effective on a variety of cottonwood hybrids (genus *Populus*). *Plant Cell Rep.* 19: 315-320.

Hare, P.D. and Chua, N.H., (2002) Excision of selectable marker genes from transgenic plants. *Nat. Biotechnol.* 20: 575 - 580.

Hay, I., Lachance, D., Aderkas, P., Charest, P.J. and Von Aderkas, P. (1994) Transient chimeric gene expression in pollen of five conifer species following microparticle bombardment. *Can. J. For. Res.* 24: 2417-2423.

Herschbach, C., Jouanin, L. and Rennenberg, H. (1998) Overexpression of gamma-glutamylcysteine synthetase, but not of glutathione synthetase, elevates glutathione allocation in the phloem of transgenic poplar trees. *Plant Cell Physiol.* 39: 447-451.

Ho, C.K., Chang, S.H., Tsay, J.Y., Tsai, C.J., Chiang, V.L. and Chen, Z.Z. (1998) *Agrobacterium tumefaciens*-mediated transformation of *Eucalyptus camaldulensis* and production of transgenic plants. *Plant Cell Rep.* 17: 675-680.

Hu, W.J., Harding, S.A., Lung, J., Popko, J.L., Ralph, J., Stokke, D.D., Tsai, C.J. and Chiang, V.L. (1999) Repression of lignin biosynthesis promotes cellulose accumulation and growth in transgenic trees. *Nat. Biotechnol.* 17: 808-812.

James, R.R., Difazio, S.P., Brunner, A.M. and Strauss, S.H. (1998) Environmental effects of genetically engineered woody biomass crops. *Biomass and Bioenergy* 14: 403-414.

Kajita, S., Osakabe, K., Katayama, Y., Kawai, S., Matsumoto, Y., Hata, K. and Morohoshi, N. (1994) *Agrobacterium*-mediated transformation of poplar using a disarmed binary vector and the overexpression of a specific member of a family of poplar peroxidase genes in transgenic poplar cell. *Plant Sci.* 103: 231-239.

Kasuga, M., Liu, Q., Miura, S., Yamaguchi-Shinozaki, K. and Shinozaki, K. (1999) Improving plant drought, salt, and freezing tolerance by gene transfer of a single stress-inducible transcription factor. *Nat. Biotechnol.* 17: 287-91.

Kesseli, R., Ochoa, O., and Michelmore, R. (1991) Variation at RFLP loci in Lactuca spp. and origin of cultivated lettuce (L. sativa). *Genome* 34: 430-436.

Kidwell, K.K., Austin, D.F., and Osborn, T.C. (1994) RFLP evaluation of nine Medicago accessions representing the original germplasm sources for North American alfalfa cultivars. *Crop Sci.* 34: 230-236.

King G.J., Tartarini S., Brown, L., Gennari, F. and Sansavini, S. (1999) Introgression of the Vf source of scab resistance and distribution of linked marker alleles within the *Malus* gene pool. *Theor. App. Genet.* 99: 1039-1046.

Krauss, S.L. (2000) Patterns of mating in *Persoonia mollis* (Proteaceae) revealed by an analysis of paternity using AFLP: implications for conservation. *Aust. J. Bot.* 48: 349-356.

Kumar, A. and Gurumurthi, K. (2000) Genetic divergence studies on clonal performance of *Casuarina equisetifolia. Silvae Genetica* 49: 57-60.

Kumar, A. (1995) *Genetic Improvement of Casuarina equisetifolia.* Ph. D thesis, Forest Research Institute (Deemed University), Dehradun, India.

Lakshmi, M., Parani, M., Ram, N. and Parida, A. (2000) Molecular phylogeny of mangroves VI. Intraspecific genetic variation in mangrove species *Excoecaria agallocha* L. (Euphorbiaceae). *Genome.* 43: 110-115.

Loaiza-Figueroa, F. and Weeden, N.F. (1991) Effects of seed increase procedures on isozyme polymorphism in *Allium. FAO/IBPGR Plant Genetic Resources Newsletter* 83/84: 1-3.

Loos, B.P. (1994) Allozyme differentiation of European populations and cultivars of Lolium perenne L., and the relation to ecogeographical factors. *Euphytica* 80: 49-57.

Maguire, T.L., Saenger, P., Baverstock, P. and Henry, R. (2000) Microsatellite analysis of genetic structure in the mangrove species Avicennia marina (Forsk.) Vierh. (Avicenniaceae). *Mol. Ecol.* 9: 1853-1862.

Martinussen, I., Junttila, O. and Twell, D. (1994). Optimi zation of transient gene expression in pollen of Norway spruce (*Picea abies*) by particle acceleration. *Physiol. Plant.* 92: 412-416.

Mathish, N.V., Tripathi, S.B. and Gurumurthi, K. (2001) DNA- Fingerprint database management using Microsoft Access- a simple strategy to corroborate fingerprints of clones. *Plant Cell Biotechnol. Mol. Biol.* 2: 119-124.

Mayer, M.S., Tullu, A., Simon, C.J., Kumar, J., Kaiser, W.J., Kraft, J.M. and Muehlbauer, F.J. (1997) Development of a DNA marker for *Fusarium* wilt resistance in chickpea. *Crop Sci.* 37 : 1625-1629.

McCown, B.H., McCabe, D.E., Russell, D.R., Robinson, D.J., Barton, K.A. and Raffa, K.F. (1991) Stable transformation of *Populus* and incorporation of pest resistance by electric discharge particle acceleration. *Plant Cell Rep.* 9: 590-594.

Mouradov, A., Sawbridge, T., Hamdorf, B., Glassick, T., Murphy, L., Marla, S., Yang, Y., Teasdale, R.D. and Drew, R.A. (1998) Genetic engineering of reproductive sterility in *Pinus radiata. Acta-Hort.* 461: 417-423.

Muluvi, G.M., Sprent, J.I., Soranzo, N., Provan, J., Odee, D., Folkard, G., McNicol, J.W. and Powell, W. (1999) Amplified fragment length polymorphism (AFLP) analysis of genetic variation in *Moringa oleifera* Lam. *Mol. Ecol.* 8: 463-470.

Parzies, H.K., Spoor, W. and Ennos, R.A. (2000) Genetic diversity of barley landrace accessions (*Hordeum vulgare* ssp *vulgare*) conserved for different lengths of time in *ex situ* gene banks. *Heredity* 84: 476-486.

Pilate, G., Guiney, E., Holt, K., Petit-Conil, M., Lapierre, C., Leplé, J-C., Pollet, B., Mila, I., Webster, E. A., Marstorp, H.G., Hopkins, D. W., Jouanin, L., Boerjan, W., Schuch, W., Cornu, D. and Halpin, C. (2002) Field and pulping performances of transgenic trees with altered lignification. *Nat. Biotechnol.* 20: 607- 612.

Powell, W., Morgante, M., Andre, C., Hanafey, M., Vogel, J., Tingey, S. and Rafalski, A, (1996) The comparison of RFLP, RAPD, AFLP and SSR (microsatellite) markers for germplasm analysis. *Mol. Breed.* 2: 225-238.

Punja, Z.K., (2001) Genetic engineering of plants to enhance resistance to fungal pathogens- a review of progress and future prospects. *Can. J. Plant Pathol.* 23: 216-235.

Rothmann, W.H., Meilan, R.,Sheppard, L.A., Brunner, A.M., Skinner, J.S., Ma, C., Cheng, S., Jouanin, L., Pilate, G. and Strauss, S.H. (2000) Diverseeffects of overexpressionof *LEAFY* and *PTLF*, a poplar (*Populus*) homolog of *LEAFY/FLORICAULA*, in transgenic poplar and *Arabidopsis*. *The Plant J.* 22 : 235-245.

Russell, J.R., Fuller, J.D., Macaulay, M., Hatz, B.G., Jahoor, A., Powell, W. and Waugh, R. (1997) Direct comparison of levels of genetic variation among barley accessions detected by RFLPs, AFLPs, SSRs and RAPDs. *Theor. App. Genet.* 95: 714-722.

Schittenhelm, S., Gladis, T. and Rao, V.R. (1997) Efficiency of various insects in germplasm regeneration of carrot, onion and turnip rape accessions. *Plant Breed.* 116 : 369-375.

Shah, D.M., Rommens, C.M.T. and Beachy, R.N. (1995) Resistance to diseases and insects in transgenic plants: progress and applications to agriculture. *Trends Biotechnol.* 13: 362-368.

Shani, Z., Levi, I., Mansfield, W., Roiz, L., Dekel, M. and Shoseyov, O. (2001) *Modulation of Wood Fibres and Studies on the Mode of Action of Cellulose-Binding Domains (CBDs)*. Paper presented at the Trees Biotechnology in the New Millennium IUFRO/ Molecular Biology of Trees Meeting, July 22-27, Stevenson, Washington, USA.

Steiner, A.M., Ruckenbauer, P. and Goecke, E. (1997) Maintenance in genebanks, a case study: contaminations observed in the Nurnberg oats of 1831. *Genet. Resour. Crop Evol.* 44: 533-538.

Tanksley, S.D. and Jones, R.A. (1981) Application of alcohol dehydrogenase allozymes in testing the genetic purity of F1 hybrids of tomato. *Hort. Sci.* 16 : 179-181.

Tohme, J., Gonzalez, D.O., Beebe, S. and Duque, M.C. (1996) AFLP analysis of gene pools of a wild bean core collection. *Crop Sci.* 36: 1375-1384.

Tripathi, S.B., Mathish, N.V. and Gurumurthi, K. (2001) *AFLP analysis in genetic fidelity testing in micropropagated Eucalyptus*.Presented in the International Symposium on Recent Advances in Biological Sciences, Oct 11 & 12, 2001 at K.S.R. college of Arts and Sciences, Tiruchengode, India.

Tsai, C.J., Popko, J.L., Mielke, M.R., Hu, W.J., Podila, G.K. and Chiang, V.L., (1998). Suppression of O-methyl transferase gene by homologous sense transgene in quaking aspen causes red-brown wood phenotypes. *Plant Physiol.* 117: 101-112.

Tuominen, H., Sitbon, F., Jacobsson, C., Sandberg, G., Olsson, O. and Sundberg, B., (1995) Altered growthand wood characteristics in transgenic hybrid aspen expressing *Agrobacterium tumefaciens* T-DNA indoleacetic acid-biosynthetic genes. *Plant Physiol.* 109: 1179-1189.

Tzfira, T., Vinocur, B., Altman, A. and Vainstein, A. (1998) Transgenic *Populus tremula*: root development, root-borne bud regeneration and *in vitro* propagation efficiency. *Trees Struct. Funct.* 12: 464-471.

Van Doorsselaere, J., Baucher, M., Chognot, E., Chabbert, B., Tollier, M.T., Petit-Conil, M., Leplé, J.C., Pilate, G., Cornu, D., Monties, B., van Montagu, M., Inzé, D., Boerjan, W. and Jouanin, L. (1995) A novel lignin in poplar trees with a reduced caffeic acid/ferulic acid O-methyltransferase activity. *Plant J.* 8: 855-864.

Varghese, M., Nicodemus, A, Nagarajan, B., Sasidharan, K.R., Siddappa, Bennet S.S.R. and Subramanian, K. (2000) *Seedling Seed Orchards- for Breeding Tropical Trees.* Institute of Forest Genetics and Tree Breeding, Coimbatore, India.

Virk, P.S., Zhu, J., Newbury, H.J., Bryan, G.J., Jackson, M.T. and Ford-Lloyd, B.V. (2000) Effectiveness of different classes of molecular marker for classifying and revealing variation in rice (*Oryza sativa*) germplasm. *Euphytica* 112: 275-284.

Weigel, D. and Nisson, O., (1995). A vv developmental switch sufficient for flower initiation in diverse plants. *Nature* 377: 495-500.

Yasodha, R., Kathirvel, M., Sumathi, R., Gurumurthi, K., Sunil, Archek and Nagaraju, J. (2004) Genetic analyses of Casuarinas using ISSR and FISSR markers. *Genetica* 122: 161-172.

Yibrah, H.S., Gronroos, R., Lindroth, A., Franzen, H., Clapham, D. and Arnold, S.V. (1996) *Agrobacterium rhizogenes*-mediated induction of adventitious rooting from *Pinus contorta* hypocotyls and the effect of 5-azacytidine on transgene activity. *Transgenic Res.* 5: 75-85.

□□□

Forest Biotechnology in India © 2008
Editors: S.A. Ansari, C. Narayanan & A.K. Mandal, pp.25-30
Satish Serial Publishing House, Delhi

2

Evaluation of Genetic Diversity in Teak Populations in Kerala Using DNA Markers

E.P. INDIRA, PRAMOD N. NAIR AND S. SABNA PRABHA

Kerala Forest Research Institute, Peechi 680 653, Kerala, India

ABSTRACT

Knowledge on the genetic variation existing within and between different populations is important for effective utilization and conservation of a particular species. Studies were conducted to assess the genetic diversity within and between teak populations growing in Kerala using co-dominant nuclear gene markers following Single Strand Confirmation Polymorphism (SSCP) approach. It was observed that populations from Konni were more diverse than others. Total sum of genetic diversity of teak was unequally distributed within and between populations.

Keywords : Agarose gel electrophoresis, allelic richness, gene diversity, PCR, specific primers.

INTRODUCTION

Teak (*Tectona grandis* L. f.) is an important timber-yielding tree species naturally occurring in forests of India, Myanmar, northern Thailand, Laos and Indonesia. Being a quality hardwood species, teak is widely planted both in its original area of distribution and in other countries in tropical Asia, Africa and Latin America. Teak genetic resources have been disturbed during the last 50 to 100 years through uncontrolled logging and movement of planting materials.

Studies with molecular markers have indicated that tree populations generally have high genetic diversity, which is unequally distributed within and between populations. Type of alleles and their frequency distribution can differ substantially among populations growing in different areas. This varying distribution of genetic diversity reflects the historical pattern of gene flow and the breeding system of a particular species. To understand population dynamics of teak, co-dominant DNA markers, which can reveal the actual frequency of heterozygous individuals, could be used.

The genetic diversity of teak from India, Thailand and Indonesia has been estimated by isozyme variation of provenances (Kertadikara and Prat, 1995). Overall understanding of the genetic diversity for teak is still fragmentary (Kjaer and Siegismund, 1996; Chantragoon and Szmidt, 2000; Lowe *et al.*, 2005; Nicodemus *et al.*, 2005). Hence, we started a detailed study on the genetic diversity in Indian teak using co-dominant molecular markers. This paper presents preliminary results of the above study.

MATERIALS AND METHODS

Natural teak populations were selected in Konni (Kaduvappara and Kattathi), Nilambur (Poochappara and Padukka) and Thrissur (Vazhani) forest divisions in Kerala state. From each population, 35-40 adult trees were randomly selected and young leaves were collected for DNA extraction. Total genomic DNA was extracted using modified CTAB method (Doyle and Doyle, 1987) and quantified using Ultra Lum Total Lab software.

DNA amplification with selected primers

Specific primers, designed by Dr. Hugo Volkaert (Kasetsart University, Thailand) for selective amplification of 6 loci coding for the 4 enzymes systems, viz., CAT (1 locus), G3PDH (3 loci), ADH (1 locus) and AAT (1 locus) were tested for polymorphism. Subsequently, five gene markers namely, CAT1, G3PDHcy1, G3PDHcy2, G3PDHcp and ADH2 were selected for further studies.

DNA amplifications were done in 12 μl volume containing 2 μl DNA, 0.1 μl Taq polymerase, 1.2 μl primer, 2.4 μl dNTPs and Taq buffer. DNA amplification was performed in a Programmable Thermal

Cycler (PTC - 100; MJ Research, USA) equipped with a heated lid to minimize sample fluxing. The PCR was programmed to denature DNA at 94ºC for 45 seconds, anneal DNA to primer at 50ºC for 45 seconds and to polymerise DNA at 72ºC for 1 min and 15 s. After 32 cycles, a final extension of 10 minutes was allowed. Restriction enzyme digestion of PCR products was done for all the primers, except CAT because they generated long fragments. For G3PDH cy2, HAE III was used for cutting the fragments. The sample was denatured by heating at 98°C for 10 min followed by immediate cooling on ice. The electrophoresis was performed on non-denaturing polyacylamide gels using the Single Strand DNA Polymorphism approach (Orita *et al.*, 1989). 5 ml of the aliquots were loaded on 8% non-denaturing poly acrylamide gel in Biorad vertical electrophoresis apparatus (30 cm x 50 cm x 0.5 mm; Biorad, Australia) and run in 1x TBE buffer at 4ºC inside a refrigerator at a constant current of 10 watt for 8 h. After the electrophoresis, silver staining was done to reveal the DNA fragments. The banding patterns that appeared on silver stained SSCP gel were scored.

Only clear bands were scored for further analysis. Different genotypes were identified based on the banding pattern. Both homozygous and heterozygous individuals were identified, as the markers are co-dominant. Only data of two nuclear gene markers viz., CAT and G3PDHcy2 are presented here.

The data were analysed using the software FSTAT Version 2.9.3 (Jérôme Goudet, 2002). The gene diversity and allelic richness within and between different populations were estimated and a comparison was made between populations.

RESULTS AND DISCUSSION

There was marked variation between individuals within population for the loci analysed (Fig. 1). The estimates of genetic parameters are given in Table 1. From the results it can be seen that Konni populations are rich in gene diversity and allelic richness. Thrissur population exhibited allelic richness and presence of specific alleles. Konni populations were found to be more homozygous for the locus G3PDH cy2 than other populations. Even within the Konni forest division, Kattathi was found to be more diverse and rich in alleles than Kadervappara. One specific allele for CAT1 was observed in Kattathi of Konni (allele 11) as well as in Vazhani of Trichur forest Division (allele 10). In Vazhani, one allele was observed for G3PDH cy2 (allele 7). Gene conservation is largely a problem for low-frequency alleles. Hence, teak populations with many specific alleles may be conserved, ensuring the continuation of evolutionary processes that create and maintain these genetic variations (Amaral *et al.*, 2004).

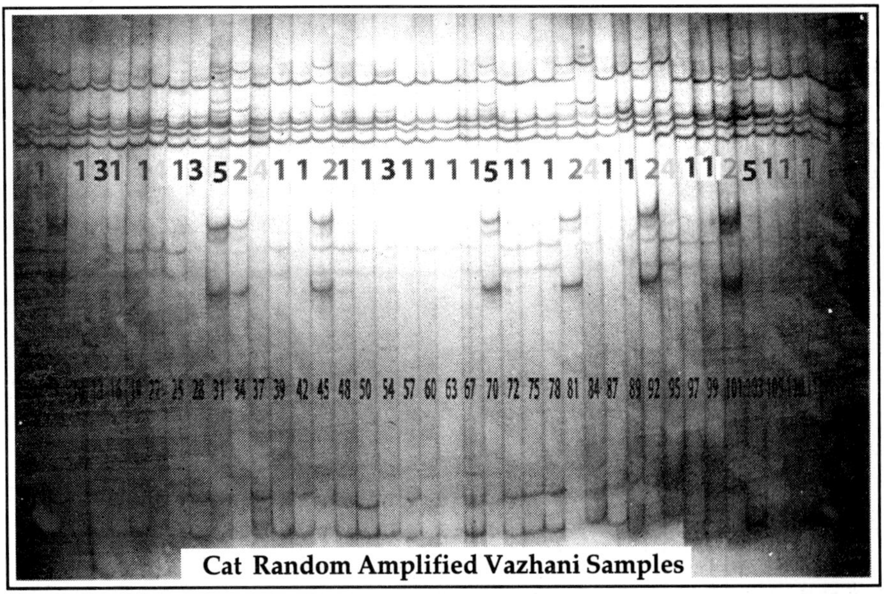

Fig. 1. SSCP gel showing diversity within population for the locus CAT.

Table 1: Gene diversity, allelic richness and heterozygosity in different teak.

Parameter	Locus	Nilambur 1 (Poochapara)	Nilambur 2 (Padukka)	Konni 1 (Kaduvappara)	Konni 2 (Kattathi)	Thrissur (Vazhani)	Mean/ Total
Gene diversity	CAT	0.566	0.534	0.658	0.772	0.500	0.650
	G3PD H cy2	0.598	0.442	0.692	0.508	0.446	0.692
No. of genotypes	CAT	5	5	6	8	7	
	Cy2	4	3	5	4	5	
Allelic richness	CAT	3.0	3.0	3.0	4.998	3.999	5.089
	Cy2	3.0	2.0	5.695	5.382	4.203	5.721
% Heterozygosity	CAT	55.5	30.8	47.5	77.1	45.0	
	Cy2	29.6	43.6	15.8	5.9	21.9	

Total gene diversity was found to be 0.671 of which 0.572 (85%) was distributed within populations which is in agreement with an earlier study (Nicodemus *et al.*, 2005) where it is reported that 78% of total diversity was distributed within populations. Our separate studies on wood density in teak provenances also showed that 78 percent of the variability is within provenances/ populations. As out crossing in teak is more than 98 percent, there would be more within population diversity. Genetic differentiation was found to be 14.75 percent.

Since within population diversity was found to be very high, it is advisable to assemble trees from the same locality to establish seed orchards rather than from different places as suggested by Nicodemus *et al.* (2005). This can avoid non synchronous flowering among clones as well as non flowering due unfavorable site.

Analysis of more teak populations is still continuing, and the results are expected to give a clear picture of Indian teak with regard to the genetic diversity.

ACKNOWLEDGEMENT

The authors would like to express their sincere gratitude to Dr. Hugo Volkaert, Kasetsart University, Thailand for providing the primers and for the help in standardizing the methods. Authors also acknowledge the financial assistance by the European Union to carry out the studies. Thanks are due to The Director, KFRI for encouragement and facilities provided.

References

Amaral, W., Kjær, E., Yanchuk, A and Graudal, L. (2004) Research Needs. *In: Forest genetic resources conservation and management: Overview, concepts and some systematic approaches, vol. 1.*International Plant Genetic Resources Institute, pp 37- 47.

Changtragoon, S. and Szmidt, A.E. (2000) Genetic Diversity of Teak (Tectona grandis Linn. F.) in Thailand revealed by Random Amplified Polymorphic DNA (RAPD). *Paper presented at IUFRO Working Party 2.08.01 Tropical Species Breeding and Genetic Resources: Forest Genetics for the Next Millennium.* International Conference Centre, Durban, South Africa.

Doyle, J.J. and Doyle, J.L. (1987) TAB DNA extraction in plants. *Phytochem. Bull.* 19: 11-15.

Jérôme Goudet. (2002) *FSTAT version 2.9.3.2.*

Kertadikara, A.W.S. and Prat, D. (1995) Isozyme variation among teak (*Tectona grandis* L.f.) provenances. *Theor. Appl. Genet.* 90: 803-810.

Kjær, E.D. and Siegismund, H.R. (1996) Allozyme diversity in two Tanzanian and two Nicaraguan landraces of teak (*Tectona grandis* L.). *Forest Genet.* 3: 45–52.

Lowe, A., Davies, S., Volkaert, H., Cheua-ngam, J., Indira E.P., Balasundaran, M., Sudarsono, Palupi, E., Wellendorf, H., Skov, E., van der Straeten, D. and Shrestha, M. (2005) TEAKDIV - Developing know-how for the improvement and sustainable management of teak genetic resources. *In Proc. International Conference on Quality Timber products of teak from sustainable Forest Management* (eds. Bhat, K.M., Nair, K.K.N., Bhat, K.V. Muralidharan, E.M.and Sharma, J.K.) KFRI, Peechi, pp. 361- 365.

Nicodemus, A., Nagarajan, B., Narayanan, C., Varghese, M. And Subramanian, K. (2005) RAPD variation in indian teak populations and its implications for breeding and conservation. *In Proc. International Conference on Quality Timber products of teak from sustainable Forest Management* (eds. Bhat, K.M., Nair, K.K.N., Bhat, K.V. Muralidharan, E.M.and Sharma, J.K.) KFRI, Peechi, pp. 347- 351.

Orita, M., Iwahana, H., Kanazawa, H., Hayashi, K. and Sekiya, T. (1989). Detection of polymorphisms of human DNA by gel electrophoresis as single-strand conformation polymorphism. *Proc. Natl. Acad. Sci. USA* 86: 2766-2770.

❑❑❑

Forest Biotechnology in India © 2008
Editors: S.A. Ansari, C. Narayanan & A.K. Mandal, pp.31-36
Satish Serial Publishing House, Delhi

3

Impact of Human Disturbance on Genetic Diversity in Teak-An Assessment Through Nuclear Gene Markers

PRAMOD N. NAIR, E.P. INDIRA,
S. SABNA PRABHA AND R. RAJALAKSHMI

Kerala Forest Research Institute, Peechi 680 653 Kerala, India

ABSTRACT

A study was conducted to evaluate the impact of human disturbance on genetic diversity in teak populations. Paired (disturbed and undisturbed) natural teak populations were selected in four teak growing areas, Konni, Nilambur, Thrissur and Wynad, in Kerala and the genetic diversity was estimated through molecular markers using Single Strand Confirmation Polymorphism (SSCP) approach and microsatellites. The genetic diversity and allelic richness within and between populations were estimated and compared to evaluate the impact of human disturbance on genetic diversity. There is no indication of decline in diversity due to human disturbance in these areas with regard to the adult populations. However, the genetic diversity and allelic richness were comparatively less in the progeny population compared to the parent population in one of the disturbed site.

Keywords : Allelic richness, gene diversity, Konni population, Nilambur population, specific primers.

INTRODUCTION

Genetic diversity is vital for both plant and animal communities. Habitat loss and forest fragmentation is causing widespread decline in biodiversity by adversely affecting survival, reproduction and dispersal in many species (Woodwards, 2005). Most investigations emphasize the potential dangers of fragmentation in isolating patches of forest and exposing populations to loss of species diversity through founder effects, genetic drift, inbreeding, and restricted gene flow. Forest fragmentation can lead to genetic isolation of small populations that can cause inbreeding and genetic drift (Bawa, 1994). Lillesøe (1996) reports that logging of non-commercial species in a forest area in Indonesia has reduced the population size and efficiency of specific pollinators, which in turn has severely reduced the reproduction capacity of some of the commercially important *Dipterocarpus* species.

Hence, a molecular genetic analyses study was conducted at Kerala Forest Research Institute to evaluate the impact of human disturbance on genetic diversity and gene flow in teak populations in Kerala state.

MATERIALS AND METHODS

Twin populations (disturbed and undisturbed) were selected in natural teak forests in Konni, Nilambur, Thrissur and Wynad Forest Divisions in Kerala. Number of left-over stumps observed, presence of old logging roads, frequency of forest fires, presence or absence of natural regeneration, distance to inhabited areas, records of recent disturbances etc. were taken into account to classify the populations as disturbed and undisturbed forests.

In each of the populations, 35-40 adult trees were randomly selected. Young leaves were collected for DNA extraction and total genomic DNA was extracted using modified CTAB method (Doyle and Doyle, 1987). DNA samples were quantified using Ultra Lum Total Lab software. In order to maintain uniform concentration, 30 ng DNA samples were used for further studies.

DNA amplification with selected primers

Specific primers, designed by Hugo Volkaert (Kasetsart University, Thailand) for selective amplification of six loci coding for four enzyme systems viz., Cat (1 locus), G3PDH (3 loci), ADH (1 locus) and AAT (1 locus), were used in this study. These six nuclear gene markers were tested for their polymorphism. Subsequently, four gene markers namely Cat1, G3PDHcy1, G3PDHcy2, G3PDHcp and ADH2 were selected for further amplification studies.

DNA amplifications with nuclear primers and electrophoresis were carried out as described by Indira *et al* (2006). Single Strand Confirmation Polymorphism (SSCP) technique (Orita *et al.*, 1989) was adopted to detect very small variations. Only two nuclear gene markers CAT and G3PDHcy2 were analysed. The data were analysed using the software FSTAT Version 2.9.3 (Jérôme Goudet, 2002). The genetic diversity and allelic richness within and between populations were estimated and a comparison was made to evaluate the impact of human disturbance on genetic diversity.

120 parent trees and 120 progenies from Thamaravellachal, a disturbed area in Thrissur forest division of Kerala, were further analysed using highly polymorphic microsatellite markers. Eight microsatellite primers viz., AC01, AC28, AG04, AG14, AG16, ADHMS, CPIMS and AC44 were selected for amplification (India *et al.,* 2006). DNA amplifications were same as that for nuclear gene markers, except for annealing temperature, which was 48^{0}C instead of 50^{0}C, DNA polymerization at 72^{0}C for 1 min and final extension of 3 min instead of 10 min. The amplified products were separated on 4.5% denaturing poly acrylamide gel at 90 watts constant current for 90 min in 1x TBE buffer. The gels were silver stained and photographed. Based on the banding pattern alleles were identified manually. Data was analysed using FSTAT.

RESULTS AND DISCUSSION

The results obtained using nuclear gene markers are given in Table 1. There was no significant difference between disturbed and undisturbed populations with regard to gene diversity, number of alleles, allelic richness and reduction in heterozygosity. But there was marked difference between Konni and Nilambur populations with regard to all the above factors except homozygosity. Konni populations were rich in all the above factors (Table 1, Fig. 1).

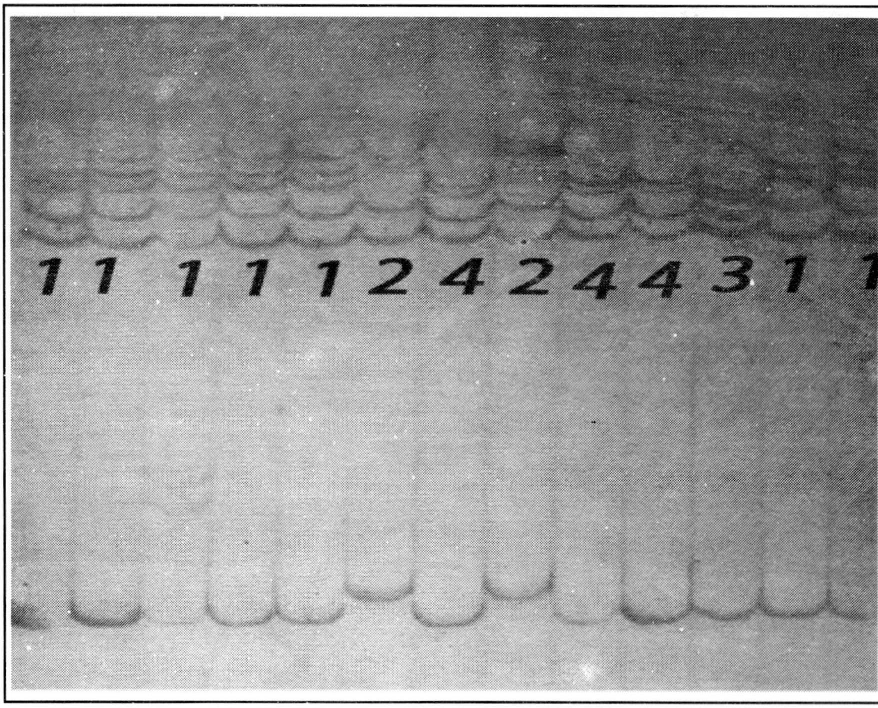

Fig. 1. SSCP Gel showing variations within Konni population for the locus G3PDHcy2.

Table 1: Gene diversity, allelic richness and homozygosity in disturbed and undisturbed natural teak populations based on nuclear gene markers

		Teak population			
		Nilambur		Konni	
Parameter	**Locus**	**Undisturbed**	**Disturbed**	**Undisturbed**	**Disturbed**
Gene diversity	CAT	0.566	0.534	0.658	0.772
	G3PDH cy2	0.598	0.442	0.692	0.508
No. of alleles	CAT	3	3	3	5
	Cy2	3	2	6	6
Allelic richness	CAT	3.0	3.0	3.0	4.998
	Cy2	3.0	2.0	5.695	5.382
Fis (reduction in heterozygosity)		0.268	0.343	0.508	0.352
Homozygosity	CAT	44.5	69.2	52.5	22.9
	Cy2	70.4	56.4	84.2	78.1

Since there was no marked difference between disturbed and undisturbed natural teak populations, the progeny population in one of the disturbed areas in Thrissur forest Division was further investigated in detail as described earlier. The genetic parameters estimated in parents and progenies in the disturbed teak population using microsatellite markers showed that there was a reduction in gene diversity, number of alleles and allelic richness in progenies compared to parent populations (Table 2). The reduction in heterozygosity also was more in progeny (0.012) compared to the parent population (0.006). Our results are commensurate with that of Woodward (2005) who reported that fragmentation only reduced genetic variation in seedlings, with no effect on adults in *Matisia ochrocalyx*. But this may be true only with a viable minimum population beyond which fragmentation could affect diversity within adult population that may vary with species. Even though more than 98 percent cross pollination is reported in this particular population (Sabna Prabha *et al.*, 2006), there was reduction in genetic diversity. Moreover, there may be a loss of specific alleles reported in some populations (Indira *et al.*, 2006) due to deforestation. In addition to diversity reduction as opined by Savolainen and Kärkäinen (1992), the genetic improvement following moderate selection is probably low, whereas severe degradation of forests may cause an important genetic impairement in terms of offspring with poor stem form and reduced commercial value.

Table 2: Gene diversity, allelic richness and heterozygosity in parents and progenies in disturbed natural teak population in Thrissur Forest Division based on microsatellite markers.

		Microsatellite markers							
		Ac 01	**Ac 28**	**Ac 44**	**Ag 04**	**Ag 14**	**Ag 16**	**ADHms**	**CPIms**
Gene diversity	Parents	0.550	0.378	0.624	0.539	0.799	0.760	0.494	0.291
	Progenies	0.471	0.254	0.520	0.290	0.715	0.745	0.49 8	0.341
Number of alleles present	Parents	8	5	7	5	7	6	2	3
	Progenies	6	5	6	5	5	6	2	2
Allelic richness	Parents	7.91	5.00	6.94	5.00	6.94	5.99	2.00	2.93
	Progenies	6.00	5.00	6.00	4.99	5.00	5.88	2.00	2.00
Reduction in heterozygosity	Parents	Mean = 0.006							
	Progenies	Mean = 0.012							

The results presented in this paper are indicative and need more investigation. More teak populations are being analysed at KFRI, the result of which may give a better picture of the impact of human disturbance on teak diversity.

ACKNOWLEDGEMENT

The authors would like to express their sincere gratitude to Dr. Hugo Volkaert, Kasetsart University, Thailand for providing the primers and for the help in standardizing the methods. Authors also acknowledge the financial assistance by the European Union to carry out the studies. Thanks are due to The Director, KFRI for encouragement and facilities provided.

References

Bawa, K.S. (1994) Effects of Deforestation and Forest Fragmentation on Genetic Diversity in Tropical Tree Populations. *In: Proc. of International Symposium on Genetic Conservation and seed production of tropical forest tree seed* (eds. Drysdale, R.M, John, S.E.T. and Yapa, A.C.) ASEAN-Canadian Seed Centre Project. Muaklek, Thailand, pp.10-16.

Doyle, J.J. and Doyle, J.L. (1987) A rapid DNA isolation procedure from small quantities of fresh leaf tissues. *Phytochem. Bull.* 19:11-15.

Indira, E.P., Nair, Pramod N. and Sabna Prabha, S. (2006) Evaluation of Genetic diversity in teak populations in Kerala using DNA markers. *In this proceedings.*

Jérôme Goudet (2002) *FSTAT version* 2.9.3.2.

Lillesøe, J.P.B. (1996). *Dipterocarps- A review of Literature of Relevance to Tree Seed Source Programmes with Emphasis on the Species in Malesia. Second, slightly revised edition.* Ministry of Forestry, Indonesia Tree Seed Source Development Project. PT. Ardes Perdana/Danagro Adviser A/S. Indonesia.

Orita, M., Iwahana, H., Kanazawa, H., Hayashi, K. and Sekiya T. (1989) Detection of polymorphisms of human DNA by gel electrophoresis as single-strand conformation polymorphism. *Proc. Natl. Acad. Sci. USA* 86: 2766-2770.

Sabna Prabha, S., Indira, E.P., Pramod N. Nair and Rajalakshmi, R. (2006) Evaluation of mating system and contemporary gene flow in teak using microsatellite markers. *In this proceedings.*

Savolainen, O. and Karkkainen K. (1992) Effect of forest management on gene pools. New For. 6: 329–345.

Woodward, C.L. (2005) Reproductive Success, Genetic Diversity, and Gene Flow in Fragmented Populations of Two under story Tree Species in Costa Rica. Ph.D.Thesis, University of Wisconsin, Madison.

□□□

Forest Biotechnology in India © 2008
Editors: S.A. Ansari, C. Narayanan & A.K. Mandal, pp.37-41
Satish Serial Publishing House, Delhi

4

AFLP Marker Analysis for Genotyping of Teak Spp.

S.A. ANSARI*, C. NARAYANAN, S.A.WALI, N.SHUKLA, R. KUMAR AND A.K. MANDAL

Genetics & Plant Propagation Division, Tropical Forest Research Institute, P.O.–RFRC, Mandla Road, Jabalpur 482 021, India

ABSTRACT

Plus trees of *Tectona grandis* and *T. hamiltoniana* were analysed using AFLP (Amplified Fragment Length Polymorphism) technique. The technique generated highly reproducible AFLP profiles or fingerprints, useful for differentiating the different plus trees of *T. grandis* as well as for distinguishing both teak species. The identity of teak plus trees got validated by generation of specific AFLP profile commensurate with discreteness of individual tree.

Keywords **:** AFLP, genotyping, Teak.

INTRODUCTION

Teak (*Tectona grandis* Linn.f.), belonging to the family Verbenaceae, enjoys the reputation of being the paragon among timber yielding tropical forest trees. Teak grows in the natural forests of India, Myanmar, Laos, Thailand and Indonesia (Tewari, 1995). Considering that selection can improve productivity and quality of forest trees, teak attracted several tree improvement programs in India since 1950s.

In the initial stages of genetic improvement programme, several outstanding phenotypically superior teak trees were identified from natural forests/plantations of different states and referred to as plus trees (Mandal and Rambabau, 2001). They were selected based on superior phenotypic characters, viz., (i) vigour (better height and girth), (ii) straight, cylindrical clear bole without branching, flutting and buttressing, (iii) narrow or compact crown with light branching (iv) reasonable amount of seed production.

The recent development of molecular markers offers quick and unambiguous means for assessment of genetic worth of individuals/ populations. Of these, amplified fragment length polymorphism (AFLP) is a versatile technique and makes use of restriction enzyme digestions followed by selective amplification of fragments involving ligation of terminal adaptor sequences and PCR primers modified by addition of two or three selective nucleotides (Vos *et al.*, 1995). It is a very powerful semi-arbitrary priming fingerprinting technique (Zabeau and Vos, 1993; Vos *et al.*, 1995). AFLP technique has been widely applied in genome mapping, fingerprinting, analyses of genetic relationships and genetic diversity. The aim of the present study is to evaluate AFLP markers for DNA fingerprinting and genome analysis of plus trees of teak as well as analysis of teak species.

MATERIALS AND METHODS

Plant material

2-3 ramets (shoot branches) of forty eight plus trees Indian teak (*Tectona grandis* L.f.) belonging to eleven states of India and one check tree of *T. hamiltoniana* Wall., an endemic species to Myanmar, were collected and planted in mist chamber, for bud break and sprout (leaf) production for DNA extraction.

DNA extraction and AFLP analysis

Genomic DNA of plus trees extracted from fresh leaves using CTAB method was subjected to conventional AFLP protocol developed by Vos *et al.* (1995). Selective amplification was done using *Eco* RI and *Mse* I primers with selective nucleotides. The amplified samples were size-fractionated on polyacrylamide gel and the fragments were detected by autoradiography.

RESULTS AND DISCUSSION

The four AFLP primers generated highly diagnostic and reproducible fingerprint profile for each plus trees of teak. The primers also generated specific profiles for the two species (Fig. 1).

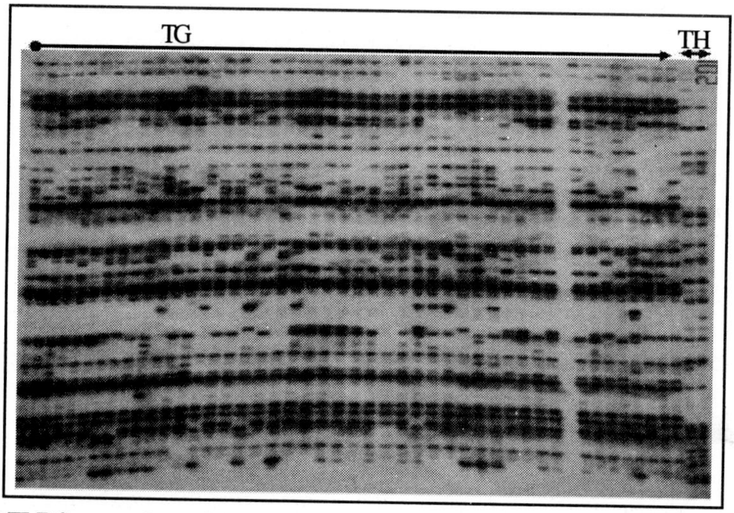

Fig. 1. AFLP fingerprints of teak (*Tectona grandis*) plus trees (TG) and *T. hamiltoniana* (TH).

Genetic variation of teak has been studied mainly using RAPD, AFLP and microsatellite markers (Nicodemus *et al.*, 2005; Parthiban *et al.*, 2003; Watanabe *et al.*, 2004; Shrestha *et al.*, 2005; Verhaegen *et al.*, 2005) and genetic uniformity of tissue culture plants using ISSR markers (Yasodha *et al.*, 2003). Here, we observed that in comparison to ten RAPD and five ISSR primers (Nicodemus *et al.*, 2003; Yasodha *et al.*, 2003), only four AFLP primers were needed to detect total sum polymorphism and distinguish teak plus trees among themselves and with that of *T. hamiltoniana* as well.

Jaccard's similarity coefficients for plus trees ranged 0.15-1.00 in AFLP analysis which was higher than those obtained for RAPD and ISSR for the same samples (Narayanan *et al.*, 2007). AFLP analysis involves restriction sites of genome which are highly conserved, whereas the RAPD and ISSR analyses make coverage of random sites or specific variable sites of the genome resulting in detection of high level of similarity in the former than the latter. Such a comparison of individual genotypes using all the three molecular markers has been performed in cashew (Archak *et al.*, 2003).

Thus, AFLP generated DNA fingerprints for the plus trees of teak can be used for maintenance of accurate identification of the above plus trees that is important in germplasm management and also maintain correct identity of plus trees in large-scale planting or breeding program.

ACKNOWLEDGEMENT

Authors are grateful to Department of Biotechnology, Ministry of Science and Technology, Government of India, New Delhi for financial support and The Energy Resources Institute, New Delhi for performing AFLP analysis.

References

Archak, S., Gaikwad, A.B., Gautam, D., Rao, E.V.V.B., Swamy, K.R.M., Karihaloo, J.L. (2003) Comparative assessment of DNA fingerprinting techniques (RAPD, ISSR and AFLP) for genetic analysis of cashew (*Anacardium occidentale* L.) accessions of India. *Genome* 46: 362-369.

Mandal, A.K. and Rambabu, N. (2003) Quantitative genetic aspects of teak improvement. In: *Genetics and Silviculture of Teak* (eds.Mandal, A.K. and Ansari, S. A.). International Book Distributors, Dehra Dun, India, pp.135-145.

Narayanan, C., Wali, S., Shukla, N., Kumar, R., Mandal, A.K. and Ansari, S.A. (2007). RAPD and ISSR markers for Imolecular Characterization of teak (*Tectona grandis*) plus trees. *J. Trop. Forest Sci.* 19 : 218-225.

Nicodemus, A., Nagarajan, B. Narayanan, C., Varghese, M. and Subramanian, K. (2005) Genetic variation in Indian teak (*Tectona grandis* L.f.) populations assayed through RAPD markers. *Indian For.* 13: 1121-1131

Parthiban, K.T., Surendran, C., Paramathma, M., Saikumar, K. (2003) Molecular characterization of teak seed sources usings RAPDs. In: *International Conference on Quality Timber Products of Teak from Sustainable Forest Management.* Kerala Forest Research Institute, India, pp. 202-209.

Shrestha, M. K., Volkaert, H. and Van Der Straeten, D. (2005) Assessment of genetic diversity in *Tectona grandis* using amplified fragment length polymorphism markers. *Can. J. For. Res.* 35: 1017-1022.

Tiwari, D.N. (1992) *A Monograph on Teak (Tectona grandis Linn.)*. International Book Distributors, Dehra Dun, India.

Verhaegen, D Ofori, D., Fofana, I., Poitel, M., and Vaillant, A. (2005) Development and characterization of microsatellite markers in *Tectona grandis* (Linn. f). *Mol. Ecol. Notes* 5: 945-947

Vos, P., Hogers, R., Bleeker, M., Reijans, M., van der Lee, T., Hornes, M., Frijters, A., Pot, J., Peleman, J., Kuiper, M. and M. Zabeau. (1995) AFLP: a new technique for DNA fingerprinting. *Nucl. Acids Res.* 23: 4407-4414.

Watanabe, A., Widyatmoko, A., Rimbawanto, A. and Shiraishi, S. (2004) Discrimination of teak (*Tectona grandis*) plus tree using selected random amplified polymorhic DNA (RAPD) markers. *J. Trol. Forest Sci.* 16: 17-24.

Yasodha, R., Sumathi, R. and Gurumurthi, K. (2003) Tissue Culture Strategies for Quality Planting Stock Production of Teak. In: *International Conference on Quality Timber Products of Teak from Sustainable Forests Management.* Kerala Forest Research Institute, Peechi, India. pp. 292-301.

Zabeau, M. and Vos, P. (1993) *Selective Restriction Fragment Amplification: A General Method for DNA Fingerprinting*. European Patent Application number: 92402629.7, Publication number EP 0534858.

□□□

Forest Biotechnology in India © 2008
Editors: S.A. Ansari, C. Narayanan & A.K. Mandal, pp.43-49
Satish Serial Publishing House, Delhi

5

Evaluation of Mating System and Contemporary Gene Flow in Teak Using Microsatellite Markers

S. SABNA PRABHA, E.P. INDIRA, PRAMOD N. NAIR AND R. RAJALAKSHMI

Kerala Forest Research Institute, Peechi 680 653, Kerala, India

ABSTRACT

Pollen flow in teak was evaluated using co-dominant microsatellite markers. The study was conducted in a natural teak forest at Thamaravellachal of Peechi-Vazhani wild life sanctuary in Kerala state. All the adult trees in the population were marked, mapped and finger printed using 8 microsatellite markers. Few seeds collected from 6 randomly selected trees were also fingerprinted. DNA fingerprint of each of the seedlings/ embryos were compared to the known seed parent and all the potential pollen parents using CERVUS software. Using the exclusion method the male parent could also be identified. Percentage of selfing and contemporary gene flow were estimated.

Keywords : DNA fingerprinting, pollen flow distance, *Tectona grandis*.

INTRODUCTION

DNA markers have indicated that tree populations generally contain a wealth of diversity, which is unequally distributed within and between populations. This distribution of diversity is naturally controlled by gene flow and the breeding system of the species. Knowledge of the genetic variation within and between populations is important for the effective utilization and conservation of species. Factors determining the level and structure of genetic variation within plant species include population density, mating system and mechanisms of gene flow (Loveless, 1992). The seed dispersal curve, together with pollen dispersal curve are important tools in estimating the minimum area that is required for conservation of a viable population of trees. Hence, the information on contemporary and long term gene flow as well as mating system in natural forests and plantations are essential to chalk out effective utilization and conservation measures. The relative contribution of pollen and seed dispersal to gene flow is not well understood in most tropical tree species. This information is important for determining the size of buffer zones and conservation areas. In teak, so far not much studies were conducted on contemporary and long term gene flow as well as migration pattern.

Microsatellite markers are used in seed orchard management in various ways. They are used in studies on mating pattern, male fertility, percent selfing, pollen dilution zone, pollen contamination from out side sources and identification of clones for mass propagation. In natural forests they are used to study the mating system and contemporary gene flow to evolve effective utilization and conservation measures.

MATERIALS AND METHODS

Plant material

The study was conducted in a natural teak population at the Peechi -Vazhani Wild Life Sanctuary (Thrissur, Kerala). 190 trees were randomly selected and marked and GPS coordinates taken. The population was mapped using the software Mapinfo professional. Fruits were also collected from 6 randomly selected fruit bearing trees (mother trees) and were put to germination. Leaf samples were collected from all the marked adult trees as well as 20 germinated progeny seedlings from each of the 6 selected mother trees. Few seeds did not germinate and hence, embryos were collected. In addition to the above, hundred fruits were randomly selected and investigated for seeds per fruit.

DNA extraction

DNA was extracted from young leaves using modified CTAB procedure (Doyle and Doyle, 1987). DNA samples were quantified using Ultra Lum Total Lab software. In order to maintain uniform concentration, 30 ng DNA samples were used for further study.

DNA Amplification With Microsatellites

Hyper variable microsatellite DNA markers were used for paternity analysis and pollen flow studies. Eight microsatellite primers, viz., AC01, AC28, AG04, AG14, AG16, ADHMS, CPIMS and AC44 (Table 1), designed and supplied by Hugo Volkaert, Kasetsart University, Thailand, were used. DNA amplifications were done in 12 µl volumes containing 2 µl of DNA, 0.1 µl of Taq polymerase, 1.2 µl of primer, 2.4 µl dNTP's and Taq buffer. DNA amplification was performed in a Programmable Thermal Cycler (PTC- 100, MJ Research, USA) equipped with a heated lid to minimize sample fluxing. The Thermal Cycler was programmed as follows: 94^{0}C for 45 seconds, 48^{0}C for 45 seconds and 72^{0}C for 1 minute. After 32 cycles, a final extension of 3 minutes was allowed. The amplified products were separated on 4.5% denaturing poly acrylamide gel at 90 watts constant current for 90 minutes in 1x TBE buffer. The gels were silver stained and photographed by EPSON Photo PC 3100Z. Based on the banding pattern in the gel, alleles were manually identified.

Table 1: Primers and number of alleles present in the population.

Primers	No. of alleles
AC01	8
AC28	6
AC 44	7
Ag04	5
Ag14	6
Ag16	6
ADH-Ms	2
CPI-Ms	2

Totally, 190 trees from the population as well as 120 progenies raised from six mother trees were fingerprinted. The genotypic fingerprint of each seedling was compared to the known seed parent and the potential pollen parents determined using the software CERVUS. Through exclusion method, by comparing each and every microsatellite band, the male parent could be identified. From this, the distance between pollen parent and seed parent could be measured. The number of pollinators in different distance classes like 1-50, 50-100, 100-150 m etc. were also computed. The percentage of selfing and crossing was also estimated.

RESULTS AND DISCUSSION

The study showed that more than 45 percent (45.32) of the fruits were seedless, 43.17 % had one seed, 7.91% had 2 seeds, 3.24 % contained 3 seeds and only 0.36% developed 4 seeds. The pollen parents of the selected progenies could be identified through CERVUS analysis. The percentage of crossing was computed, which showed that 98.3 % of the total seeds set through cross pollination and the rest by selfing. Pollen reception by a female parent was found to be from almost all direction (Fig.1). It was also found different trees acted as pollen donors. In few cases, pollens were received more than once from the same male parent. With regard to tree no.151, 15 different male parents contributed pollen to produce the 20 progenies analysed. Out of the 20 progenies, three received pollen from tree no.66, two each from trees 41, 17 and 183. Remaining 11 progenies had 11 different male parents. Even though the above population is highly disturbed, high cross pollination could possibly lead to high within population diversity. High within - population diversity in teak has already been reported by Nicodemus *et al.* (2005) and Indira *et al.* (2006). The very high out crossing rate in teak can ensure production of high quality seeds in seed orchards (Fig.1).

The studies conducted by Aldrich and Hamrick (1998), using hypervariable genetic markers in *Symphonia globulifera* in southern Costa Rica, showed that, out of 250 progenies collected from a single forest fragment, 68 percent received pollen trees in nearby pastures, and remaining from trees within the fragments. It was noted that trees left standing in pastures can actually dominate the reproduction in nearby remnant forests, creating a "genetic bottleneck". Our result showing pollination of a single tree by many surrounding trees can have important implication in teak breeding.

The pollen dispersion was found to be up to 400 m (Fig. 2) but maximum pollen flow was between 150-200 m (Table 2). Hence, pollen

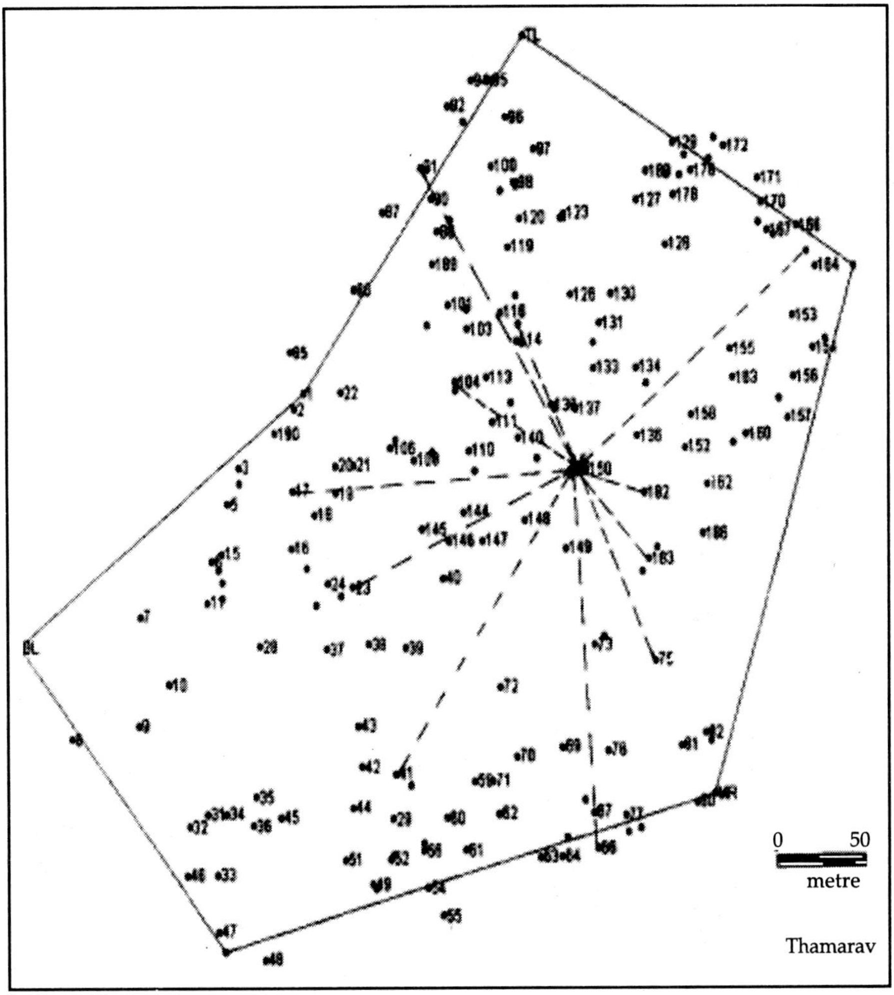

Fig. 1. Pollen flow from different paternal trees to material tree (no. 151).

dilution zone may be limited to 200 metres around seed orchards. Even if there are flowering trees in the neighbourhood, the female parents receive and accept pollen from different trees, thereby ensuring the diversity. Hence, seeds from the same tree may have different genetic constitution leading to high within population genetic diversity. Studies of historical and contemporary gene flow levels, via genetic structure surveys and parentage analyses, demonstrated that gene flow is generally extensive in both wind- and animal-pollinated forest tree species (Slavov *et al.*, 2003). Our studies with teak, an entomophilous species, support the above observation.

Table 2 : Distance between male and female parents.

Progeny ID	Material tree ID	Paternal tree ID	Distance of pollen flow (m)
1	151	104	85.37
2	151	66	175.61
3	151	165	170.73
4	151	41	180.49
5	151	66	175.61
130	100	100	Self pollinated
131	100	144	160.98
132	100	139	114.63
133	100	139	114.63
134	100	75	243.90

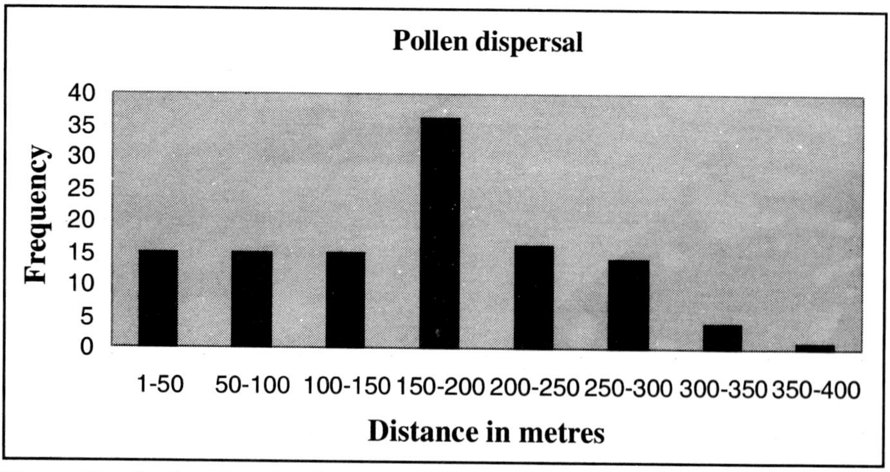

Fig. 2. Graph showing the frequency of pollen dispersal in different distance classes.

ACKNOWLEDGEMENT

The authors would like to express their sincere gratitude to Dr. Hugo Volkaert, Kasetsart University, Thailand for providing the primers and for the help in standardizing the methods. Authors also acknowledge the financial assistance by the European Union to carry out the studies. Thanks are due to The Director, KFRI for encouragement and facilities provided.

References

Aldrich, P.R. and Hamrick, J.L. (1998) Reproductive dominance of pasture trees in a fragmented tropical forest mosaic. *Science* 281: 103- 105.

Doyle, J.J. and Doyle, J.L. (1987) CTAB DNA extraction in plants. *Phytochem. Bull.* 19: 11-15.

Indira, E.P., Pramod N. Nair and Sabna Prabha, S. (2006) Evaluation of Genetic diversity in teak populations in Kerala using DNA markers. *In this proceedings.*

Loveless, M.D. (1992) Isozyme variation in tropical trees: patterns of genetic organization. *New For.* 6: 67-94.

Nicodemus, A., Nagarajan, B., Narayanan,C., Varghese, M. And Subramanian, K. (2005). RAPD variation in indian teak populations and its implications for breeding and conservation. In: *Proc. International Conference on Quality Timber products of teak from sustainable Forest Management* (eds. Bhat, K.M., Nair, K.K.N., Bhat, K.V., Muralidharan, E.M. and Sharma, J.K.) KFRI, Peechi, India, pp. 347- 351.

Slavov, G.T., DiFazio, S.P. and Strauss, S.H. (2003) Gene Flow in Forest Trees: Gene Migration Patterns and Landscape Modeling of Transgene Dispersion in hybrid Poplar. In: *Introgression from Genetically Modified Plants into Wild Relatives* (eds. den Nijs, H.C.M, Bartsch, D.and Sweet, J.) CAB International, UK., pp. 89-106.

R $(CA)_7$. Hundred per cent polymorphism was detected in *E. urophylla* species with primer UBC842, where 12 ISSR fragments were obtained. In *E. camaldulensis,* all primers amplified a comparatively large number of loci and the number of bands ranged from 10 to 24. More than 80% polymorphism was detected with all the ISSR primers except RA $(GCT)_6$. Among the primers tested, R$(CA)_7$ and R$(GCT)_6$ amplified maximum number (31) and minimum number (10) of fragments, respectively. Maximum number of polymorphic bands (27) was amplified by R$(CA)_7$ in *E. tereticornis.* Among the six species, maximum genetic diversity in terms of percent polymorphic loci was observed in *E. tereticornis* (47.27%), whereas *E. citriodora* exhibited the lowest genetic diversity (18.64%). A similar trend was observed for Nei's mean gene diversity, with minimum mean gene diversity in *E. citriodora,* and maximum mean gene diversity in *E. tereticornis* (Table 2). Nei's genetic similarities and distances between the six *Eucalyptus* species are shown in Table 3. Nei's similarity index was the highest (0.8969) between *E. camaldulensis* and *E. tereticornis,* whereas diversity was the highest between *E. citriodora* and *E. grandis.* An AMOVA was used to partition total genetic diversity among genera, populations among species within genera and within species (Table 4). The AMOVA results indicated that 53.32% ($P = 0.0127$) of the variation was attributable to differences between species; diversity among the species was insignificant (15.65%, $P > 0.02346$). Genetic diversity within different populations was 31.03% ($P = 0.03617$). The genetic relationships among the six species are shown in the dendrogram (Fig. 1). The dendrogram showed a close relationship between *E. camaldulensis* and *E. tereticornis.* It revealed wide genetic diversity among individuals of *E. tereticornis, E. camaldulensis* and E. *urophylla,* whereas a narrow genetic diversity was observed among individuals of *E. citriodora, E. grandis* and *E. pellita.* Among populations, maximum similarity was observed between two individuals of *E. grandis* with a similarity coefficient of 0.96. On the other hand, maximum diversity was observed between two individuals of *E. camaldulensis* with a similarity coefficient of 0.57 (data not shown). The dendrogram clearly illustrated wide diversity between *E. citriodora* and the other species (Fig. 1).

Species-specific markers

The study revealed twenty species-specific markers for four species, viz., *E. camaldulensis, E. citriodora, E. grandis* and *E. urophylla*

Table 2: Genetic variability within individuals of *Eucalyptus* species detected by ISSR analysis.

Species	na	ne	h	I	P (%)
E. camaldulensis	1.3955	1.1824	0.1118	0.1733	39.55
E. citriodora	1.1864	1.1088	0.0637	0.0955	18.64
E. grandis	1.2318	1.141	0.0803	0.1193	23.18
E. pellita	1.2455	1.13	0.0786	0.1197	24.55
E. tereticornis	1.4727	1.1993	0.1246	0.1958	47.27
E. urophylla	1.3091	1.1556	0.0941	0.1443	30.91

na - Observed number of alleles per locus, ne - Effective number of alleles per locus, h - diversity, I - Shannon's Information index, P - Polymorphic loci.

Table 3: Genetic similarity and genetic distance among six *Eucalyptus* species.

Species	Eca	Eci	Egr	Epe	Ete	Eur
Eca (*E. camaldulensis*)	****	0.7579	0.8487	0.8400	0.8969	0.8559
Eci (*E. citriodora*)	0.2773	****	0.7512	0.7626	0.7621	0.7651
Egr (*E. grandis*)	0.1640	0.2861	****	0.8188	0.8326	0.8285
Epe (*E. pellita*)	0.1743	0.2710	0.1999	****	0.8485	0.8620
Ete (*E. tereticornis*)	0.1088	0.2717	0.1833	0.1643	****	0.8401
Eur (*E. urophylla*)	0.1556	0.2677	0.1882	0.1484	0.1742	****

Genetic similarity (above diagonal) and genetic distance (below diagonal)

Table 4: Analysis of molecular variance (AMOVA) for 149 individuals in 6 populations using seven ISSR markers.

Source of variation	Degrees of freedom	Sum of squares	Variance components	Percentage of variation P*
Among genera	2	1489.46	5.66 (Va)	15.65>0.02346
Among species within genera	3	1441.17	19.29 (Vb)	53.32=0.01271
Within species	141	1582.36	11.22 (Vc)	31.03>/=0.03617

* After 1023 random permutation

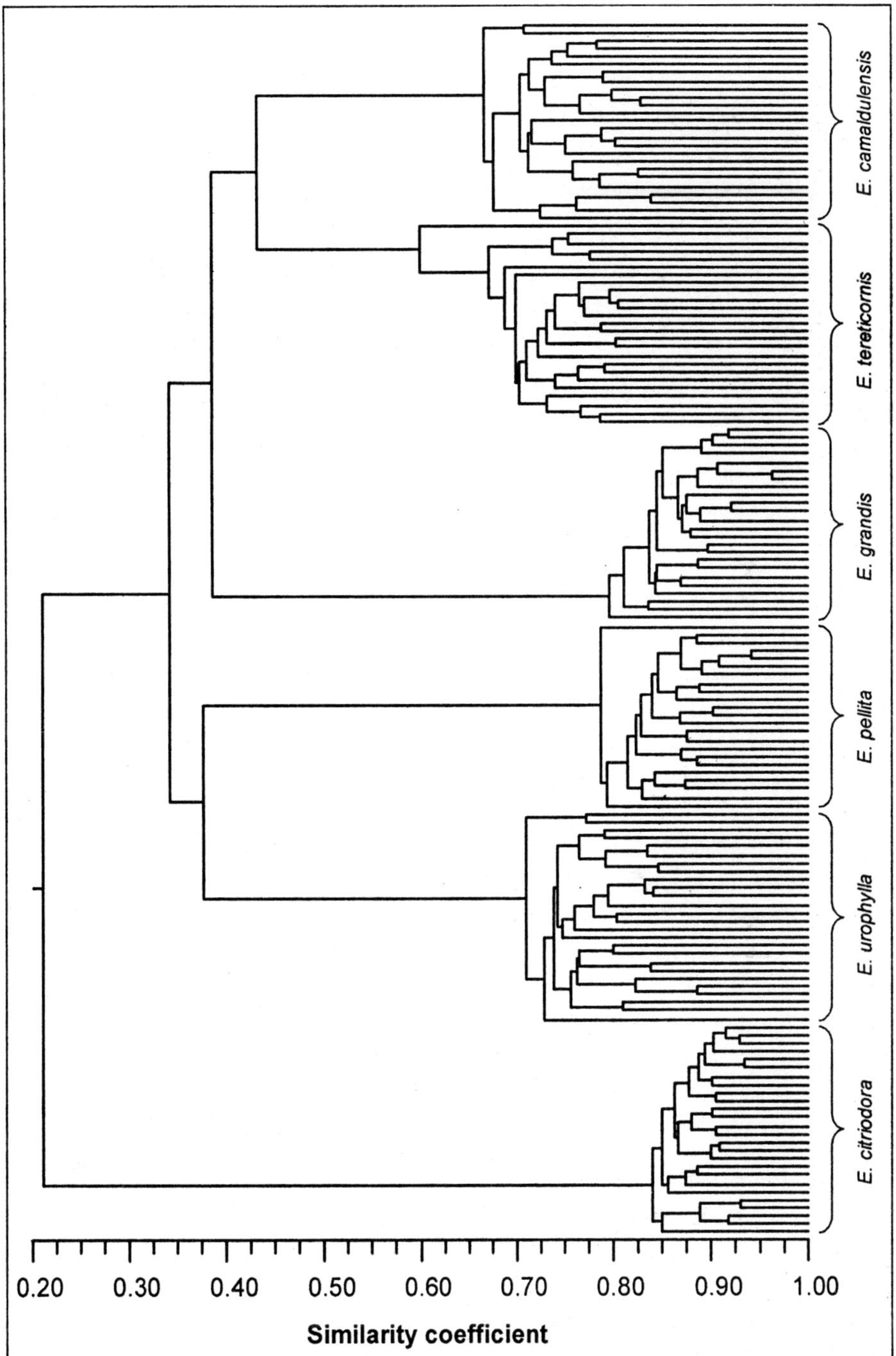

Fig. 1. Dendrogram based on similarity coefficients (Dice) among six *Eucalyptus* species obtained by ISSR markers.

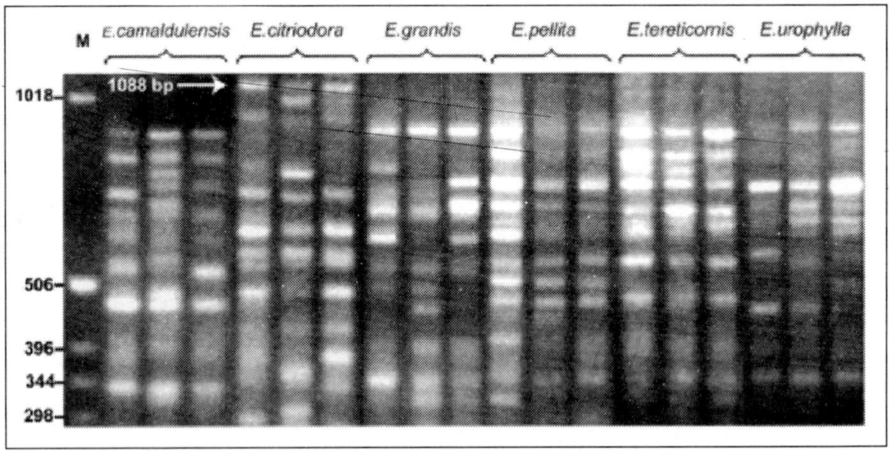

Fig. 2. ISSR profile of *Eucalyptus* species using primer UBC-842 showing a species-specific marker (MW – 1088 bp) for *E. citriodora*

(Table 5). For example, Figure 2 shows an ISSR profile of *E. citriodora* amplified with UBC842 primer with species-specific marker at 1088 bp. A maximum of eleven species-specific markers were identified for *E. citriodora* and one for *E. camaldulensis* at 813 bp. Among the specific markers identified, nine markers from *E. citriodora,* three from *E. grandis* and same number from *E. urophylla* had a frequency of 100% (Table 5). $T(GT)_9$-813 generated the only marker for *E. camaldulensis* and was present in 80% individuals. *E. citriodora* had a maximum of six exclusive species specific markers, viz., $T(GT)_9$-1150, $TA(CAG)_4$-1535, $TA(CAG)_4$-1000, $R(GCT)_6$-635, $R(GCT)_6$-490 and UBC842-1088, which were absent in rest of the species. Similarly, two exclusive markers viz., $TA (CAG)_4$-1440 and $TA(CAG)_4$-740 in *E. urophylla* and one exclusive marker viz., $TA (CAG)_4$-460 in *E. grandis,* were also identified.

DISCUSSION

In the present study, we estimated the intra-and inter-species genetic relationships of six tropical *Eucalyptus* species using ISSR markers. We observed high inter-species diversity, but low intra-species diversity. The diversity estimates for the six *Eucalyptus* spp. obtained in the present study are comparable with estimates from *Populus,* another out-crossing forest species, where 27% diversity between *P. alba* and *P. tremula* was registered by AFLP analysis (Fossati *et al.,* 2004). Nei's inter-population diversity indices and the dendrogram

Table 5: Species-specific ISSR markers in *Eucalyptus* species.

Taxon	ISSR Marker	Same Species		Other Species	
		No. of individuals	Presence (%)	No. of individuals	Presence (%)
E. camaldulensis	$T(GT)_9$-813	25	80.00	124	0
E. citriodora	$R(CA)_7$-950	26	92.31	123	0.81
	$T(GT)_9$-1150	26	100.00	123	0
	$T(GT)_9$-1018	26	100.00	123	6.50
	$T(GT)_9$-405	26	100.00	123	3.25
	$T(GT)_9$-345	26	100.00	123	6.50
	$TA(CAG)_4$-1535	26	100.00	123	0
	$TA(CAG)_4$-1000	26	100.00	123	0
	$TA(CAG)_4$-625	26	100.00	123	8.13
	$R(GCT)_6$-635	26	100.00	123	0
	$R(GCT)_6$-490	26	100.00	123	0
	UBC842-1088	26	100.00	123	0
E. grandis	$R(CA)_7$-528	24	83.33	125	4.80
	$T(GT)_9$-631	24	100.00	125	12.80
	$TA(CAG)_4$-1150	24	87.50	125	0
	$TA(CAG)_4$-577	24	100.00	125	1.60
	$TA(CAG)_4$-460	24	100.00	125	0
E. urophylla	$R(CA)_7$-390	26	100.00	123	7.32
	$TA(CAG)_4$-1440	26	100.00	123	0
	$TA(CAG)_4$-740	26	100.00	123	0

grouped *E. camaldulensis, E. grandis, E. pellita, E. tereticornis* and *E. urophylla* together and grouped *E. citriodora* separately. This grouping is concordant with the taxonomical hierarchy for the genus *Eucalyptus* wherein all the species studied, except for *E. citriodora,* belong to the subgenus Symphyomyrtus. *Eutalyptus citriodora* is classified under subgenus Corymbia (Pryor and Johnson, 1971). The genetic parameters estimated from the ISSR data indicated substantial genetic diversity within the seedling populations of each species. Genetic diversity within each *Eucalyptus* species ranged from 18.64 to

47.27% for the *E. tereticornis* population. An earlier RAPD-based study in the temperate species *E. globulus* recorded 30.7% genetic diversity among its populations (Nesbitt *et al.*, 1995). Similarly, a comparative estimation of heterozygosity using SSR markers registered greater diversity in *E. tereticornis* than in *E. globulus* (Marques *et al.*, 1998). The species-specific markers are widely used in the field of molecular taxonomy, particularly in animals (Gili *et al.*, 2004; Maier *et al.*, 2001). However, only a few reports on species-specific markers are available in plants (Scheef *et al.*, 2003; Kochieva, 1999; Kochieva *et al.*, 2004). In eucalypts, exclusive RAPD markers were reported in *E. rudis* and *E. drummondii* (Rossetto *et al.*, 1997). Similarly, RAPD markers were used to individually discriminate ten clones from different eucalypt species including *E. grandis, E. urophylla, E. grandis* x *E. urophylla, E. pellita, E. resinifera, E. tereticornis, E. paniculata, E. dunnii, E. torelliana* and *E. citriodora* (Grattapaglia *et al.*, 1992). However, ISSR, a highly informative and reproducible marker, has not been employed in eucalypts for identification of species except for development of SSRs (Van Der Nest *et al.*, 2000). In this study, species-specific ISSR markers were identified in four *Eucalyptus* species while no specific markers were observed in *E. tereticornis* and *E. pellita*. Similar problem was experienced while developing species-specific markers using RAPDs (Scheef *et al.*, 2003). This could be due to the occurrence of high divergence among the particular species populations. It is inferred that more primers should be used in future to develop diagnostic markers for these two species. The assessment of genetic variation within species of *Eucalyptus* will assist in predicting achievable genetic gain in breeding programs which may indicate the stability of progenies of interspecific crosses. The species-specific markers identified can be used for precise and rapid identification of species and putative hybrids during selection and hybridization programmes of the above *Eucalyptus* spp.

ACKNOWLEDGMENTS

The research was funded by the Department of Biotechnology, Ministry of Science and Technology, Government of India.

References

Awasthi, A.K., Nagaraja, G.M., Naik, G.V., Kanginakudru, S., Thangavelu. K., and Nagaraju, J. (2004) Genetic diversity and relationships in mulberry (genus *Morus*) as revealed by RAPD and ISSR marker assays. *BMC Genet.* Jan 10; 5 (1): 1 online version.

Boland, D.J., Brooker, M.I.H., Chippendale, G.M., Hall, N., Hyland, B.P.M., Johnston, R.D., Kleinig, D.A. and Turner, J.D. (1984) Forest Trees of Australia, Ed. 4. 687 pp. (Thomas Nelson, CSIRO, Melbourne, Australia).

Brooker, M.I.H. and Kleinig, D.A. (2001) *Field Guide to Eucalypts*.Bloomings Books. Hawthorn, Victoria.

Doyle, J.J. and Doyle, J.L. (1987) A rapid DNA isolation procedure for small quantities of fresh leaf tissue. *Photochem. Bull.* 19: 11-15.

Excoffier, L., Smouse, P.E. and Quattro, J.M. (1992) Analysis of molecular variance inferred from matrix distances among DNA haplotypes: application to human mitochondria DNA restriction data. *Genetics* 131: 479-491.

Fossati, T., Patrignani, G., Zapelli, I., Sabatti, M., Sala, F. and Castiglione, S. (2004) Development of molecular markers to assess the level of introgression of *Populus tremula* into *P. alba* natural populations. *Plant Breed.* 123: 382-385.

Gaiotto, F.A., Bramucci, M. and Grattapaglia, D. (1997) Estimation of out-crossing rate in a breeding population of *Eucalyptus urophylla* with dominant RAPD and AFLP markers. *Theor. Appl. Genet.* 95: 842-849.

Ge, X.J. and Sun, M. (1999) Reproductive biology and genetic diversity of a cryptoviviparous mangrove *Aegiceras corniculatum* (Myrsinaceae) using allozyme and intersimple sequence repeat (ISSR) analysis. *Mol. Ecol.* 8: 2061-2069.

Gemas V.J.V., Almadanim, M.C., Tenreiro, R., Martins, A. and Fevereiro, P. (2004) Genetic diversity in the Olive tree (*Olea europaea* L. subsp. *europaea*) cultivated in Portugal revealed by RAPD and ISSR markers. *Genet. Resour. Crop Evol.* 51: 501–511.

Gili, M., Monaghan, M.T. and Spaak, P. (2004) Amplified fragment length polymorphism (AFLP) reveals species-specific markers in the *Daphnia galeata-hyalina* species complex. *Hydrobiologia* 526: 63-71.

Grattapaglia, D., Malley, D.O. and Sederoff R.(1992). Multiple applications of RAPD markers to genetic analysis in Eucalyptus sp. In: *Proceedings of the IUFRO Meeting "Resolving Tropical Forest Resources Concerns through Tree Improvement, Gene Conservation and Domestication of New Species"* (IUFRO WP S2.02–08), pp. 436 – 450.

Hantula, J., Dusabenygasani, M. and Hamelin, R.C. (1996) Random amplified microsatellites (RAMS) - a novel method for characterizing genetic variation within fungi. *Eur. J. For. Path.* 26:15-166.

Kallarackal, J. and Somen, C.K. (1997) An ecophysiological evaluation of the suitability of *Eucalyptus grandis* for planting in the tropics. *For. Ecol. Manag.* 95: 53-61.

Kallarackal, J., Somen, C.K. and Rajesh, N. (2002) Studies on Water Use of Six Tropical Eucalypt Species in Kerala. In : *Recent eucalypt research in India* (eds. Bagchi., S.K. Varghese, M. and Siddappa), pp 94-115.

Kimura, M. and Crow, J. (1964) The number of alleles that can be maintained in a finite population. *Genetics.* 49: 725-738.

Kochieva, E.Z.(1999).Molecular markers of potato and tomato species and cultivars genome. Pan European Network on the Genetic Indicators of Biodiversity (PEN/GIB): Description of Diversity and Exchange of Information by Network Web, Florence, 25-27 October.

Kochieva, E.Z., Ryzhova, N.N., Molkanova, O.I., Kudryavtsev, A.M., Upelniek, A.P.and Okuneva, I.B. (2004). The genus *Syringa*: Molecular markers of species and cultivars. *Russ. J. Genet.* 40: 30–32.

Lai, J.A., Yang, W.C. and Hsiao, J.Y. (2001) An assessment of genetic relationships in cultivated tea clones and native wild tea in Taiwan using RAPD and ISSR markers. *Bot. Bull. Acad. Sin.* 42: 93-100.

Maier, E., Tollrian, R. and Nurnberger, B. (2001) Development of species-specific markers in an organism with endosymbionts: microsatellites in the scleractinian coral *Seriatopora hystrix*. *Mol. Ecol. Notes* 1:157-159.

Marques, C.M., Araujo, J.A., Ferreira, J.G., Whetten, R., O'Malley, D.M., Liu, B.H. and Sederoff, R. (1998) AFLP genetic maps of *Eucalyptus globulus* and *E. tereticornis*. *Theor. Appl. Genet.* 96: 727-737.

Nagarajan, B., Varghese, M., Ravi, N., Nicodemus, A. and Sivakumar, V. (2002) Controlled pollination and full sib production of *Eucalyptus tereticornis* and *Eucalyptus camaldulensis*. Proceedings of International conference on eucalypt productivity: a synthesis of the physiological, environmental, genetic and silvicutural determinants of growth. Hobart, Tasmania. 10-15 November 2002. pp. 41-43.

Nei, M. (1972) Genetic distance between populations. *Am. Nat.* 106 : 283-292.

Nei, M. (1973) Analysis of gene diversity in subdivided populations. *Proc. Nat. Acad. Sci. (USA)* 70 : 3321-3323.

Nei, M. and Li., W.H. (1979) Mathematical models for studying genetic variation in terms of restriction endonucleases. *Proc. Natl. Acad. Sci. (USA)* 76 : 5269-5273.

Nesbitt, K.A., Potts, B.M. Vaillancourt, R.E. West, A.K. and Reid,J.B. (1995). Partitioning and distribution of RAPD variation in a forest tree species, *Eucalyptus globulus* (Myrtaceae). *Heredity* 74:628-637.

Poltri, S.N.M., Zelener, N., Traverso, J.R., Gelid, P. and Hopp, H.E. (2003) Selection of a seed orchard of *Eucalyptus dunnii* based on diversity criteria calculated using molecular markers. *Tree Physiol.* 23: 625-632.

Pryor, L.D. and Johnson, L.A.S. (1971) *A Classification of the Eucalyptus*. Australian National University Press, Canberra.

Rohlf, F.J. (1992) NTSYS-pc. Numerical taxonomy and multivariate analysis system, Version 1.80.

Rossetto, M., Lucarotti., F., Hopper, S.D. and Dixon, K.W. (1997) DNA fingerprinting of *Eucalyptus graniticola*: a critically endangered relict species or a rare hybrid? *Heredity* 79: 310 - 318.

Scheef, E.A., Casler, M.D. and Jung, G. (2003) Development of species-specific SCAR markers in Bent grass. *Crop Sci.* 43: 345-349.

Van Der Nest, M.A., Steenkamp, E.T., Wingfield, B.D. and Wingfield, M.J. (2000) Development of simple sequence repeat (SSR) markers in *Eucalyptus* from amplified inter-simple sequence repeats (ISSR). *Plant Breed.* 119: 433-436.

Yasodha, R., Kathirvel, M., Sumathi, R., Gurumurthi, K., Archak, Sunil and Nagaraju, J.(2004). Genetic analyses of casuarinas using ISSR and FISSR markers. *Genetica* 122: 161-172.

Yeh, F.C., Yang, R.C., Boyle, T.J.B., Ye, Z.H. and Mao, J.X. (1997) *POPGENE, the User-Friendly Shoftware for Population Genetic Analysis*. Edmonton, Molecular Biology and Biotechnology Centre, University of Alberta, Canada.

Zietkiewicz, E., Rafalski, A. and Labuda, D. (1994) Genome fingerprinting by simple sequence repeats (SSR)-anchored polymerase chain reaction amplification. *Genomics* 20: 176-183.

□□□

Forest Biotechnology in India © 2008
Editors: S.A. Ansari, C. Narayanan & A.K. Mandal, pp.65-74
Satish Serial Publishing House, Delhi

7

Mining of Functional Information from *Eucalyptus* Genome to Develop Allele - Specific DNA Markers

R. YASODHA, R. SUMATHI, P. CHEZHIAN, R. KAMALAKANNAN, M. VARGHESE AND M. GHOSH

Division of Plant Biotechnology, Institute of Forest Genetics and Tree Breeding, Coimbatore-641 002, India

ABSTRACT

Eucalyptus is an important pulp-wood species and many genomics program are underway resulting in the accumulation of voluminous genomic and expressed DNA sequences in public databases. The expressed sequence tags (ESTs) available in the databases can be mined for analysis of structural and functional genes. They provide wealth of information on simple sequence repeats and single nucleotide polymorphism available in the transcribed genes. In this study, *in silico* analysis of 3481 sequences representing partial and full-length mRNA from *Eucalyptus* species for their use in developing SSRs or microsatellites were carried out. A total of 355 EST - SSRs were identified from 300 SSR containing ESTs. Motif size of 6, 5, 5 and 4 for di, tri, tetra, and penta nucleotide respectively were considered in locating the microsatellites. The average frequency of identified SSRs was 14.0. The dinucleotide repeats were the most abundant among the di, tri and tetra nucleotide motifs and accounted for 51 % of eucalypts genome. Primer designing analysis showed that 206 sequences with SSRs had sufficient flanking regions for polymerase chain reaction (PCR) primer synthesis. Evaluation of the usefulness of these SSRs were done using SSR primers designed from the sequences representing cinnamyl CoA reductase gene of *E.globulus* and *E.gunnii* and caffeoyl CoA 3-O-methyl transferase gene of *E. globulus* in sixteen provenances of *E. tereticornis*. The results showed that EST derived SSRs can generate polymorphic markers as all the primers showed allelic diversity among the provenances.

Keywords: EST-SSR, expressed sequenced tags, *Eucalyptus tereticornis*, simple sequence repeat.

INTRODUCTION

Microsatellites, or simple sequence repeats (SSRs) are short tandem repeats dispersed evenly throughout the genome and markers reveal high level of allelic polymorphisms, codominant mode of inheritance, transferable across species and analytically simple. SSRs have been widely applied in various areas of forestry including studies of genome variation, genome mapping, integration of physical and genetic maps and comparative genome analyses (Echt *et al.*, 1999).

In *Eucalyptus*, more than 60 SSR primer pairs targeting $(AG)_n$ and $(AC)_n$ repeats, designated as EMBRA, were initially developed (Brondani *et al.*, 1998, 2002). Five primer pairs (FMRSA) were developed through ISSR enrichment technique and their transferability across species was documented. Subsequently, twelve primer pairs, designated EMCRC, for the amplification of SSRs, were isolated (Steane *et al.*, 2001). Additionally about 30 primer pairs were developed using $(CA)_n$ and $(CAG)_n$ genomic-enriched library (Ottewel *et al.*, 2005). EMBRA primers have been widely used for individual identification, development of genetic linkage maps for inter and intra specific crosses and also employed in QTL analysis of various commercially important traits (Brondani *et al.*, 1998, 2002; Marques *et al.*, 2002; Rungis *et al.*, 2004; Kirst *et al.*, 2005).

The transferability of SSR markers between species has made comparative genome analysis possible for eucalypt species, where SSRs constitute anchor points for specific regions of the genome of different species (Marques *et al.*, 2002). Comparative mapping also facilitates cloning of orthologous genes by identifying cross-reference genes (Varshney *et al.*, 2005a). However, the number of SSR primer pairs developed so for in eucalypts is lesser (about 100 SSRs) than other forest tree species like *Pinus* and *Populus* (Kirst *et al.*, 2005). More number of SSR markers is essential to have high resolution saturated genetic map and to develop comparative mapping of other eucalypts. Previously, SSRs were developed from SSR enriched libraries, however, recently, an alternative source comes from data mining sequence information from the publicly available EST, genes and cDNA clonal sequence databases (Mahalakshmi *et al.*, 2002; Qureshi *et al.*, 2004).

Recently, Varshney *et al.* (2005b) reviewed the importance of EST-SSRs including their various applications and potential uses in plant genetics and breeding. Analysis of *Eucalyptus* EST databases revealed one SSR for every four sequences (Ceresini *et al.*, 2005) or one SSR per 2.5 kb sequence (Rabello *et al.*, 2005). SSRs primers developed from such databases were used for genome analysis of barley (Thiel *et al.*, 2003), wheat (Gupta *et al.*, 2003) and cotton (Han *et al.*, 2004). However, there is no such information available for eucalypts. In this study, the expressed sequences available in GenBank were used to investigate the type and distribution of repeat motifs in eucalypts and genetic diversity of eucalypts provenances.

MATERIALS AND METHODS

Plant material and DNA isolation

Seeds of sixteen provenances were obtained from Australian tree seed center, CSIRO, Australia. The seedlings were raised and planted in Panampally Research Station, Kerala, India, and used for the present study. Methods of DNA isolation has been described elsewhere (Balasaravanan *et al.*, 2005).

Source of sequences

All *Eucalyptus* ESTs, partial and full-length mRNA sequences (as on October 2004), were downloaded from NCBI\Pub Med\Nucleotide (http://www.ncbi.nim.nih.gov) using *Eucalyptus* and "mRNA" as keywords for limits of organisms and molecules respectively. The sequences were downloaded for each species in FASTA format and saved in individual files for each species. A total of 3385 EST sequences and 96 partial and full length 5'mRNA sequences representing various species and tissues were downloaded from the database. The details for each *Eucalyptus* species are given in (Table 1).

Cluster analysis

EST sequences of less than 150 bp were not considered and sequences larger than 750 bp were clipped at their 3'side to preclude

Table 1: Summary of DNA sequences used for SSR analysis

Species	No. of sequences downloaded		No. of sequences >150 bp		Singletons	Contigs
	EST	CN	EST	CN		
E. grandis	1574	18	12	0	1425	60
E.globulus	149	53	43	2	87	7
E.gunnii	487	11	62	0	394	14
E. tereticornis	1131	0	1	0	1016	53
E.grandis x *nitens*	44	1	0	0	38	3
E. camaldulensis	0	8	0	0	0	0
E.occidentalis	0	3	0	0	0	0
E.perriana	0	2	0	0	0	0
Total	3385	96	118	2	2960	137

the inclusion of low quality sequences in further analysis. ESTs downloaded from public databases were clustered using an EST assembly program CAP3 (Huang and Madan, 1999) with the criteria of at least "30 bases overlap" and "75% percent identity" between one end of a sequence and one end of the other sequences to reduce redundancy and to develop consensus sequences. The contigs and singletons generated by CAP3 were saved in FASTA formatted files for SSR detection. Core nucleotide sequences were directly used without any modification.

Detection of SSRs

The perfect SSRs were identified using Simple Sequence Repeats Identification Tool (Temnykh *et al.*, 2001) with the criteria of minimum number of repeats as "12" and maximum motif size as "5". The repeat lengths of 6, 5, 5 and 4 for di, tri, tetra, and penta nucleotides respectively, were selected for the study.

Designing of primers

Once SSR containing sequences were identified flanking primers of size 18-27 nucleotides in length were designed using PRIMER3 (Rozen and Skaletsky, 1998) based on the criteria of 55 % optimum

GC content, 60°C optimum melting temperature, and absence of secondary structure. In the present study, primers were synthesized for cinnamyl Co-A reductase gene sequence of *E. gunnii* (AJ132750), and *E. globulus* (AY656822) and caffeoyl Co-A methyl transferase (CCOMT) mRNA sequence of *E. globulus* (AF046122) to examine the allelic variations existing among the provenances.

PCR conditions and electrophoresis

PCR was carried out in 10 µl reactions consisting of 1 x PCR buffer, 1.5 µM $MgCl_2$, 200 µM dNTPs, 250 nm of primer, 0.5 U *Taq* polymerase (Genei, Bangalore) and 30 ng of genomic DNA. All fragments were amplified using PTC-200 Thermal Cycler (MJ research Inc., USA) with the temperature profile of 3 min initial denaturation at 94°C followed by 35 cycles of denaturation at 94°C for 30s, annealing at 55°C for 30s and extension at 72°C for 30s followed by a final extension step at 72°C for 5 min.

PCR products were separated on denaturing polyacrylamide gels consisting 5% polyacrylamide (AA:BIS = 9.5:0.5) and 7 M urea in 1 x TBE buffer. PCR reactions were mixed with equal volume of loading buffer (formamide, 0.8M EDTA and traces of xylene cyanol and bromophenol blue), denatured at 95°C for 5 min and snap cooled on ice. Afterwards, samples were loaded on preheated Sequi-Gen GT sequencing gel (Bio Rad, USA), which were run at 1800 V for 2-3 h. After electrophoresis, the fragments were visualized by silver staining.The gel was documented using digital camera and scoring was done by visual inspection.

RESULTS AND DISCUSSION

Single-pass sequencing of cDNA clones is considered as the best approach for gene identification in many economically important species including forest trees. Targeting the transcribed genes can be utilized in various ways including the development of simple sequence repeats (SSR) and single nucleotide polymorphism (SNP) markers. EST-SSR markers have also been widely used as anchor points in comparative genome mapping among species (Yu *et al.*, 2004; La Rota *et al.*, 2005). In case of industrially important tree like eucalypts, genome sequencing is in progress for few species (Nehra *et al.*, 2005) and hence the SSR polymorphisms generated from ESTs could be extended to other less characterized species. In eucalypts large research projects like "FORESTs" and "Genolyptus" have been established specifically for EST sequencing and identification (Poke *et al.*, 2005). However, these EST databases are not publicly accessible. In this study, 3385 EST and 96 partial or full-length mRNA sequences from various species of *Eucalyptus* spp. were downloaded from GenBank, and after the removal of sequences with less than 150bp, 3267 EST sequences and 94 mRNA sequences were used for further analysis of microsatellites.

Clustering of 3267 EST sequences were performed to reduce the redundant sequences and to create consensus sequences from overlapping regions. Of various eucalypt ESTs analyzed, 2960 singletons and 137 consensus sequences totaling to 3097 non-redundant sequences, were generated. The SSRIT detected 300 SSR sequences (14%) from a total of 3191 sequences, of which 149 SSRs were present either in 5′ or 3′ end. Only 206 SSRs had sufficient flanking regions to design primers. The list of primer sequences synthesized for this study is shown in Table 2. The average frequency of SSR motif in the eucalyptus sequence collection was 11.1%. Similar studies conducted in eucalypts detected 25.5% (Rabello *et al.*, 2005) and 29 % (Ceresini *et al*, 2005) of SSR containing ESTs; however these studies included mononucleotides as well as compound repeats. Cereal genome analysis showed an average of 3.2% ESTs which consisted of SSRs.

Table 2: Frequency and distribution of SSR motifs in *Eucalyptus* species

Species	No. of sequences	No. of SSR-ESTs	% of SSR-ESTs	No. of SSRs	No. of motifs				% of motifs	
					DN	TN	TTN	PN	DN	TN
E. camaldulensis	8	3	37.5	3	1	2	0	0	33	67
E. grandis	1503	176	11.7	215	109	94	5	7	51	44
E. globulus	145	11	7.5	14	9	5	0	0	64	36
E. gunnii	419	21	5.0	21	14	7	0	0	67	34
E. tereticornis	1069	89	8.3	102	41	56	3	2	40	55
E. occidentalis	4	0	0	0	0	0	0	0	0	0
E. perriana	2	0	0	0	0	0	0	0	0	0
E. grandis x *nitens*	41	0	0	0	0	0	0	0	0	0
Total	3191	300	14[a]	355	174	164	8	9	51[a]	47[a]

[a] Average of percentages

The occurrence of individual types of repeat motifs in expressed sequence tags also varied depending upon the species. The di- and tri-nucleotide motifs in eucalypts were 51 and 47 %, respectively with few representations of tetra and penta-nucleotide motifs. Similar results were reported in eucalypts (Ceresini *et al.*, 2005; Rabello *et al.*, 2005). However, tri-nucleotide repeats were found to be the most common, followed by either di-nucleotide or tetra-nucleotide repeats in most of the reports (Varshney *et al.*, 2005 a,b).

Among the di-nucleotide motifs the most abundant was TC/AG (96.6%) and other types of dimers were rarely found in all the species of *Eucalyptus*. Similar distribution was noticed in most of the plant species (Mahalakshmi *et al.*, 2002; Ceresini *et al.*, 2005; Rabello *et al.*, 2005), but in the case of animals AC/TG was found to be the most abundant (Ju *et al.*, 2005). Among the tri-nucleotide repeats CCT (10.4%), CTC (7.3%), CGG (6.7%) and TCC (6.7%) were the most common. In many plant species similar occurrence was reported (Thiel *et al.*, 2003; Ceresini *et al.*, 2005, Rabello *et al.*, 2005), except for species like *Medicago* (Mahalakshmi *et al.*, 2002) and wheat.

The variation in length and location of an SSR motif were also considered for its usefulness as a marker (Ju *et al.*, 2005). The length of the eucalyptus microsatellite ranged from 12-50 bases with an average of 18.24 ± 6.39 bases. Maximum number of SSRs contained 12-20 bases followed by 22-30 bases with very few occurrences of more than 30 bases. However in the FORESTs database the average size of the SSRs was found to be 22.42 bp (Ceresini *et al*, 2005).

To evaluate the usefulness of EST-SSRs as polymorphic markers across species and to examine the allelic diversity of genes involved in lignin biosynthetic pathway, primers were synthesized to amplify the genomic DNA of 16 provenances of *E. tereticornis*. All the three primers generated PCR products of expected size. The average number of amplified products was 5.7 and all the bands were polymorphic. Inter-specific transferability was high for the primers showing that significant portion of the markers can be transferred to related species. EST-SSR markers have been proven to be associated with or responsible for a target trait. Recently, wheat EST-SSR primers were used for candidate gene mapping (Gao *et al.*, 2004). Work is in progress to sequence the PCR products to check the sequence homology and isolate the orthologous genes from *E. tereticornis*.

In eucalypts, the occurrence and characteristics of genomic microsatellites was first reported in *E. nitens* by Byrne *et al.* (1995). Subsequently, several microsattelite-based linkage maps have been developed (Brondani *et al.*, 2002). *Eucalyptus* being an important source for pulp-wood, many genomic research programs are underway for genetic mapping, QTL mapping, and physical mapping. However, they are restricted to few temperate species. Hence, EST-SSR markers generated in the present study would gain significant role in marker assisted breeding as well as comparative mapping of tropical eucalypts.

References

Brondani, R.P.V., Brondani, C. and Grattapaglia, D. (2002) Towards a genus-wide reference linkage map for *Eucalyptus* based exclusively on highly informative microsatellite markers. *Mol. Genet. Genomics* 267: 338-347.

Brondani, R.P.V., Brondani, C., Tarchini, R. and Grattapaglia, D. (1998) Development, characterization and mapping of microsatellite markers in *Eucalyptus grandis* and *E. urophylla*. *Theor. Appl. Genet.* 97: 816-827.

Byrne, M., Marques-Garcia, M.I., Uren, T., Smith, D.S. and Moran, G.F. (1996) Conservation and genetic diversity of microsatelliteloci in the genus *Eucalyptus*. *Aust. J. Bot.* 44: 331-341.

Ceresini, P.C., Silva CL.S.P., Missio, R.F., Souza, E.C., Fischer, C.N., Guillherme, I.R., Gregorio, I. da Silva E.H.T., Cicarelli, R.M.B., da Silva M.T.A., Garcia, J.F., Avelor, G.A., Neto, L.R.P., Marcon, A.R., Junior, M.B. and Marini, D.C. (2005) Satellypus: Analysis and database of microsatellites from ESTs of *Eucalyptus*. *Gen. Mol. Biol.* 28: 589-600.

Echt, C.S., Vendramim, G.G., Nelson, C.D. and Marquardt, P. (1999) Microsatellite DNA as shared genetic markers among conifer species. *Can. J. For. Res.* 29: 365-371.

Gao, L.F., Jing, R.L., Huo, N.X., Li, Y., Li, X.P., Zhou, R.H., Chang, X.P., Tang, J.F., Ma, Z.Y. and Jia, J.Z. (2004) One hundred and one new microsatellite loci derived from ESTs (EST-SSRs) in bread wheat. *Theor. Appl. Genet.* 108: 1392-400.

Gupta, P.K., Rustogi, S., Sharma, S., Singh R., Kumar, N. and Balyan, H.S. (2003) Transferable EST- SSR markers for the study of polymorphism and genetic diversity in bread wheat. *Mol. Genet. Genomics* 270: 315–323.

Han, Z.G., Guo, W.Z., Song, X.L., Zhang, T.Z. (2004) Genetic mapping of EST-derived microsatellites from the diploid *Gossypium arboreum* in allotetraploid cotton. *Mol. Genet. Genomics* 272: 308–327.

Huang, X. and Madan, A. (1999) CAP3: A DNA Sequence Assembly Program. *Genome Res.* 9: 868-877.

Ju, Z., Wells, M.C., Kazianis, S., Rains, J.D., and Walter, R.B. (2005) An in silico Mining for Simple Sequence Repeats from Expressed Sequence Tags of *Zebrafish,*Medaka, Fundulus, and Xiphophorus. *In Silico Biol.* 5: 439-463.

Kirst, M., Cordeiro, G.D., Rezende, S.P. and Grattapaglia, D. (2005) Power of Microsatellite Markers for Fingerprinting and Parentage Analysis in *Eucalyptus grandis* Breeding Populations. *J. Hered* 96: 1–6.

La Rota, M., Kantety, R.V., Yu, J.K. and Sorrells, M.E. (2005) Nonrandom distribution and frequencies of genomic and EST-derived microsatellite markers in rice, wheat, and barley. *BMC Genomics* 6: 23 (doi: 10.1186/1471-2164-6-23).

Mahalakshmi, V., Aparana, P., Ramadevi, S. and Ortiz, R. (2002) Genomic sequence derived simple sequence repeat markers - Case study with *Medicago* spp. *Electronic J. Biotechnol.* 5: 233-242.

Marques, C.M., Brondani, R.P.V., Grattapaglia, D. and Sederoff, R. (2002) Conservation and synteny of SSR loci and QTLs for vegetative propagation in four *Eucalyptus* species. *Theor. Appl. Genet.* 105 : 474–478.

Nehra, N.S., Becwar, M.R., Rottmann, W.H., Pearson, L., Chowdhury, K., Chang, S., Wilde, H.D., Kodrzycki, R.J., Zhang, C., Gause, K.C., Parks, D.W., and Hinchee, M.A. (2005). Forest Biotechnology: Innovative Methods, Emerging Opportunities. *In Vitro Cell Dev. Biol. Plant* 41: 701-717.

Ottewell, K.M., Donnellan, S.C., Moran, G.F. and Paton, D.C. (2005) multiplexed microsatellite markers for the genetic analysis of *Eucalyptus leucoxylon* (Myrtaceae) and their utility for ecological and breeding studies in other *Eucalyptus* Species. *J. Hered.* 96: 445–451.

Poke, F.S., Vaillancourt, R.E., Potts, B.M. and Reid, J.B. (2005) Genomic research in *Eucalyptus. Genetica* 125: 79-101

Qureshi, S.N., Saha, S., Kantety, R.V. and Jenkins, J.N. (2004) EST-SSR: A New class of genetic markers in cotton. *J. Cotton Sci.* 8: 112-123.

Rabello, E., de Souza, A.N., Saito, D. and Tsai, S.M. (2005) *In silico* characterization of microsatellites in *Eucalyptus* spp.: Abundance, length variation and transposan associations. *Gen. Mol. Biol.* 28: 582-588.

Rozen, S. and Skaletsky, H.J. (1998) Primer 3. Code available at http://wwwgenome.wi.mit.edu/genome_software/other/primer3.html.

Rungis, D., Berube, Y., Zhang, J., Ralph, S., Ritland, C.E., Ellis, B.E., Douglas, C., Bohlmann, J. and Ritland, K. (2004) Robust simple sequence repeat markers for spruce (*Picea* spp.) from expressed sequence tags. *Theor. Appl. Genet.* 109: 1283–1294.

Steane, D.A., Vaillancourt, R.E., Russell, J., Powell, W., Marshall, D. and Potts, B.M. (2001) Development and Characterisation of Microsatellite Loci in *Eucalyptus globulus* (Myrtaceae). *Silva. Genet.* 50: 89-91.

Temnykh, S., DeClerck, G., Lukashova, A., Lipovich, L., Cartinhour, S., McCouch, S. (2001) Computational and experimental analysis of microsatellites in rice (Oryza sativa L.): Frequency, length variation, transposon associations, and genetic marker potential.

Thiel, T., Michalek, V. and Graner, A.(2003). Exploiting EST data-bases for the development and characterization of gene-derived SSR-markers in barley (*Hordeum vulgare* L.). *Theor. Appl. Genet.* 106: 411–422

Varshney, R.K., Sigmund, R., Borner, A., Korzun, V., Stein, N., Sorrells, M.E., Langridge, P. and Graner, A. (2005a) Interspecific transferability and comparative mapping of barley EST-SSR markers in wheat, rye and rice. *Pl. Sci.* 168: 195–202.

Varshney, R.K., Graner, A. and Sorrells, M.E. (2005b). Genic microsatellite markers in plants. *Trends Biotechnol.* 23: 48-55.

Yu, J.K., La Rota,M., Kantety, R.V., Sorrells, M.E. (2004) EST derived SSR markers for comparative mapping in wheat and rice. *Mol. Gen. Genomics* 271: 742-751.

❑❑❑

Forest Biotechnology in India © 2008
Editors: S.A. Ansari, C. Narayanan & A.K. Mandal, pp.75-87
Satish Serial Publishing House, Delhi

8

RAPD Markers for Genetic Diversity in *Eucalyptus Tereticornis* Sm. of Australian and Papua New Guinea Origin

H.S.GINWAL AND SHALINI SINGH MAURYA

Division of Genetics and Tree Propagation, Forest Research Institute, P.O. - I.P.E., Kaulagarh Road, Dehradun-248195, Uttarakhand, India

ABSTRACT

Thirteen provenances from Australia and Papua New Guinea and two seed sources from India of *Eucalyptus tereticornis* with a latitudinal range of 8^0 57′ N to 36^0 49′ N, longitude 76^0 04′ E to 152^0 58′ E within an altitudinal range of 40 m to 762 m, were evaluated for genetic diversity using the RAPD assay. Twenty random decamer primers were used to assess genetic variation among 15 different populations. A total of 223 fragments were generated out of which 188 amplicons were polymorphic. All primers produced albert variable polymorphic amplification products. The similarity coefficient between the provenances was calculated. UPGMA based clustering method were employed to construct the phylogenetic tree using Nei's genetic distance matrix. The value of similarity coefficient ranged from 0.44 to 0.70 and the value of genetic distance varied from 0.25 to 0.35. The dendrogram revealed high level of variation. Three distinct clusters were identified, which contained six provenances each in first two cluster and two provenances in the third cluster. Highest similarity was detected between Walsh River, Queensland and Mysore Gum. The population Burdekin River, Queensland did not cluster with any other population and hence found to be the most distinct one. Irregular changes in genetic distances among provenances suggested that the variation was not clinal.

Keywords : Genetic variation, provenance, RAPD, similarity index, polymorphism, UPGMA cluster.

INTRODUCTION

Eucalyptus tereticornis Sm. commonly known as forest red gum is native to Papua New Guinea and Australia. In Papua New Guinea, the species grows between 6° and 10° latitudes, at elevations of 0 to 800 m. In Australia, it grows between 15° and 38° latitudes, from Victoria through New South Wales, and in northern Queensland at elevations of 30 to 1000 m. *E. tereticornis* is a fast-growing tree that can reach 30 - 45 m in height and 1 - 2 m in diameter. The species grows in open forests or as scattered trees in alluvial plains and along streams, including brackish waters. It grows better in deep, well-drained, light-textured, neutral, or slightly acid soils (Eldridge *et al.* 1994). *E. tereticornis* has been introduced to several countries in the tropics and subtropics for production of wood for fuel, poles, construction and pulp (Zobel *et al.*, 1987; Evans, 1992). The ability to grow even under adverse environmental conditions coupled with fast growth rate makes it ideally suited to fill the ever-enlarging gap between the demand and supply for timber, fuel, etc.

Populations of *Eucalyptus* spp. in India are highly inbred and over exploited through intensive selection and multiplication for commercial plantations (Ginwal *et al.*, 2005). Lack of sufficient genetic variability and intensive selection from insufficient base population are some of the important reasons for low productivity of *Eucalyptus* plantations in India as compared to other countries. In order to broaden the genetic base of *Eucalyptus* for further selection and improvement, germplasm of *Eucalyptus* spp. has been introduced for time and again to India from its natural seeds sources.

Study of genetic variability is the basic pre-requisite for genetic improvement and conservation of genetic resources with molecular marker systems becoming indispensable in this pursuit. Morphological markers have limited use in identification of elite genotypes in trees as visual inspection is not a reliable method. In contrast, DNA markers are highly reliable and most preferred choice for germplam characterization. Random amplified polymorphic DNA (RAPD) technique (Williams *et al.*, 1990) is highly successful to study genetic diversity in many plant genera such as mango (Schnell *et al.*, 1995), *Eucalyptus* (Keil and Griffin, 1994), Himalayan poplar (Rajgopal *et al.*, 2000), cacao (Whitkus *et al.*, 1998), spruce (Scheepers *et al.*, 1997), *Populus* spp. (Castiglione *et al.*, 1993).

In the present study, thirteen seed sources of *E. tereticornis* of Australian and Papua New Guinea origin, introduced to India under evaluation trials for improving productivity, were assessed for the genetic variability through RAPD analysis. Individuals raised from two local sources viz., FRI-5 (a putative hybrid of E. *camaldulensis* x *E. tereticornis)* and Mysore gum *(E. tereticornis)* were also included for comparision. The above study was specifically undertaken to analyse the genetic base of the above germplasm of *E. tereticornis* from which further selections would be made under short rotation forestry programmes.

MATERIAL AND METHODS

Plant material

Leaf material collected from the existing provenance trial of *Eucalyptus tereticornis* established at FRI, Dehradun, from seeds obtained from Australian Seed Centre, CSIRO Australia, was used. As control, leaf material collected from the offspring of natural hybrid FRI-5 (*E. camaldulensis* x *E. tereticornis*), and Mysore gum (*E. tereticornis*), were used. The details of seed sources and their geographical locations are given in Table 1.

Table 1 : Detail of *E. tereticornis* provenances used in the study.

S.No	Provenance/ Seed lot No.	Origin			
		Locality	Lat. (° ')	Long. (° ')	Altitude (m)
1	13399	Oro bay to Emo, PNG	08 57	148 28	200
2	13418	Sirinumu Sogeri Plat, PNG	09 30	147 26	580
3	17762	Warwick, QLD	28 15	152 05	450
4	17768	Yurammie, SF, NSW	36 49	149 45	170

Contd...

Table 1. Contd...

5	17770	Buckenbowra SF, NSW	35 40	150 07	110
6	18732	Selection flat SF559, NSW	29 10	152 58	40
7	19315	Credition SF, QLD	21 7	148 31	730
8	20468	Cardwell, QLD	18 10	145 58	84
9	20469	Mitchell River MT Molloy, QLD	16 44	145 20	390
10	20470	Mill stream archer creek, QLD	17 39	145 21	670
11	20471	Helenvale, QLD	15 48	145 15	325
12	20472	Walsh River, QLD	17 20	145 18	762
13	20474	Burdekin River, QLD	19 48	146 04	291
14	Mysore gum	Dehradun, India	30 30	76 04	640
15	FRI-5	Dehradun, India	30 30	76 04	640

DNA Extraction

Genomic DNA was extracted from young leaves using the protocol described by Stange *et al.* (1998) with slight modification. Leaf material (250 mg) was ground to fine powder in liquid N_2 using a mortar and pestle and suspended in 1 ml of extraction buffer (2% CTAB, 1.42 M NaCl, 20 mM EDTA, 100mM Tris HCl, pH 8.0, 2% w/v PVP40, 5mM ascorbic acid, and freshly added β-mercaptoethanol) in 2 ml micro-centrifuge tubes. The homogenate was incubated at 60°C for 30 min. After incubation, 500 μl of chloroform: isoamyl alcohol (24:1) was added to the homogenate, mixed gently to form an emulsion and centrifuged at 14000 rpm for 10 min. The top aqueous phase was transferred to a fresh tube, and DNA was precipitated by addition of equal volume of cold iso-propanol followed by centrifugation at 10,000 rpm for 10 min. The DNA pellet was washed with wash buffer (76% ethanol, 10 mM ammonium acetate), air-dried and further treated with 5 μl RNase A (10 mgL^{-1}). After centrifugation, the pellet was washed with cold 70% ethanol, dried and resuspended in 100 μl of TE buffer (10mM Tris HCl; 1mM EDTA; pH 8.0).

RAPD Analysis

The amplification reactions were performed in a total volume of 25 µl containing 2.5 µl of 10 x *Taq* buffer (Fermentas) (750mM Tris-HCL, pH 8.8 at 25 ° C, 200 mM $(NH_4)_2SO_4$, 0.1 % Tween 20), 50mM $MgCl_2$, 5mM each of dATP, dCTP, dGTP and dTTP (Sigma), 10 µl of random decamer primers, 1 unit of *Taq* DNA Polymerase (Fermentas) and five nanogram of template DNA. Amplification was carried out in a thermal cycler (BIO-RAD). The first cycle was programmed as - denaturation for 1 min. at 94° C; annealing for 1 min. at 37° C and extension for 1 min. at 72° C. Subsequently there were 41 cycles each of denaturation for 45 sec. at 94° C, annealing for 1 min. at 37° C, primer extension for 1 min. at 72° C. An additional extension at 72° C for 10 min was performed to facilitate complete extension.

The amplified products were separated on 1.5 % agarose gels using 1 x TBE buffer at 3 Vcm^{-1} for 5 h, stained with ethidium bromide (0.5 $mgml^{-1}$) and visualized using gel documentation system (UVP). The bands were scored using Gene Profiler Software (Scanalytics, Inc. USA). Amplification with each primer was repeated at least twice in order to ensure reproducibility and only consistent bands were considered for further analysis. A total of thirty-five decamer primers already identified by Mosseler *et al.* (1992) and supplied by Operon (Qiagen, USA) were screened. Finally, twenty primers were selected based on reproducibility and polymorphic nature of bands.

Statistical Analysis

Each amplified product was scored for its presence or absence across all individuals. Co-migrating bands were considered to represent the same locus and thus treated as the same band. Pair wise comparison of the different sources was employed to calculate the similarity index. UPGMA (Unweighted Pair Group Method Analysis) clustering method was employed to construct the phylogenetic tree based on Nei's genetic distance matrix (Nei and Li, 1979) using Gene Profiler Software. The dendrogram was created using TREECON (Scanalytics, Inc.,USA) program and was graphically represented as a phenetic tree with bootstrap values, which provide a confidence level for each branch point and indicate the actual number of times the branch point occurred.

RESULTS

For effective reproducibility of RAPD assay, the concentrations of various reaction components and parameters of PCR amplification cycles were standardized. A total of twenty different RAPD primers (Table 2) was used to evaluate the level of genetic variation between fifteen provenances of *E. tereticornis*. RAPD assay revealed considerable variation in the level of polymorphism among the provenances of the *E. tereticornis*. The RAPD primers amplified varying number of markers and demonstrated a comparable degree of polymorphism across the provenances. Amplification of total genomic DNA from different sources produced a total of 223 fragments, out of which 188 were polymorphic. Amplification products (bands) obtained per primer ranged from 5 to 19, with an average of 11 per primer. Percent polymorphism ranged from 60% (OPA-04) to 100% (Mosseller-119), with an average of 84%. The amplified fragments ranged from 400 bp to 1800 bp in size (Table 2).

Table 2 : List of primers sequences with their percent polymorphism.

S. No.	Name of primer	Base sequence (5'-3 ')	Number of amplified fragments	Number of polymorphic fragments	Percent polymorp-hism
1	Mosseler-119	ATT GGG CGA T	7	7	100
2	Mosseler-29	CCG GCC TTA C	5	4	80
3	Mosseler-31	CCG GCC TTC C	16	14	88
4	Mosseler-33	CCG GCT GGA A	16	13	81
5	Mosseler-132	AGG GAT CTC C	7	5	71
6	Mosseler-147	GTG CGT CCT C	17	15	88
7	Mosseler-156	GCC TGG TTG C	13	11	85
8	Mosseler-169	ACG ACG TAG G	10	9	90
9	Mosseler-170	ATC TCT CCT G	8	5	63
10	Mosseler-182	GTT CTC GTG T	9	8	89
11	Mosseler-184	CAA ACG GCA C	12	10	83
12	Mosseler-186	GTG CGT CGC T	14	12	86
13	Mosseler-188	GCT GGA CAT C	10	9	90
14	Mosseler-191	CGA TGG CTT T	9	8	89
15	Mosseler-198	GCA GGA CTG C	9	8	89
16	Mosseler-199	GCT CCC CCA C	13	12	92
17	OPA-02	TGC CGA GCT G	11	10	91
18	OPA-03	AGT CAG CCA C	19	15	79
19	OPA-04	AAT CGG GCT G	10	6	60
20	OPA-10	GTG ATC GCA G	8	7	88

The RAPD banding patterns obtained using primer OPA02 is shown in Figure 1. With this primer as many as 11 bands were amplified in each sample of which 10 were polymorphic. The amplified fragments ranged from 320 bp to 1700 bp in size.

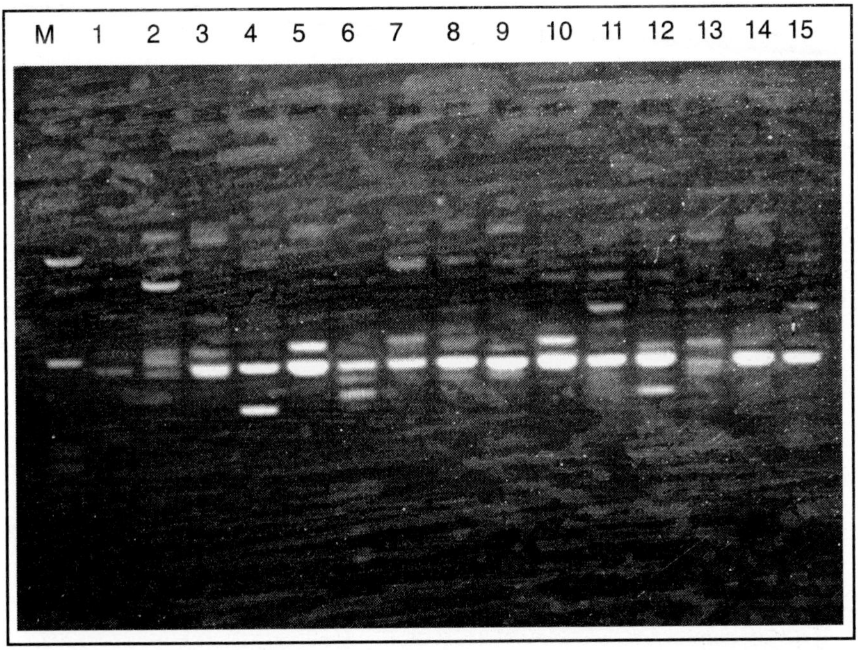

Fig. 1. RAPD profiles of different *E. tereticornis* provenances generated by primer OPA-02. Lane M, molecular weight marker (PhiX 174 DNA/*Hae* III digest); lanes 1-15, *E. tereticornis* provenances (numbers represent serial numbers of provenances given in Table 1).

The genetic similarity matrix of all pair wise combination of genotypes is presented in Table 3. Similarity coefficient values ranged between 0.44 and 0.70 indicating suggesting a wide genetic base of the fifteen sources taken in the experiment. The highest value of the index was found between Walsh River and Mysore Gum (Dehradun) followed by Yurammie, SF, NSW and Mitchell River, QLD suggesting their similar nature. The lowest value of the index was obtained among Burdekin River, QLD and Oro bay to Emo, which suggest these sources are genetically distant. The Oro bay to Emo (PNG) was noticed as the most divergent population on the basis of average similarity index. This population revealed a considerable genetic distance with Mysore gum (Dehradun, India), Mill Stream Archer Creek (QLD) and Sirinumu Sogeri Plat (PNG) in comparison to other populations.

The UPGMA-based dendrogram, constructed using Nei and Li genetic distance matrix, revealed three distinct clusters (Figure 2). The values of genetic distance varied from 0.25 to 0.35. The first cluster was found to comprise six provenances/sources, viz., Mysore gum (Dehradun, India), Walsh River, (QLD), Selection flat SF559 (NSW), Cardwel (QLD), FRI-5 and Credition SF (QLD). Within this cluster, highest similarity was observed between Mysore gum and Walsh River and between Cardwell and FRI-5.

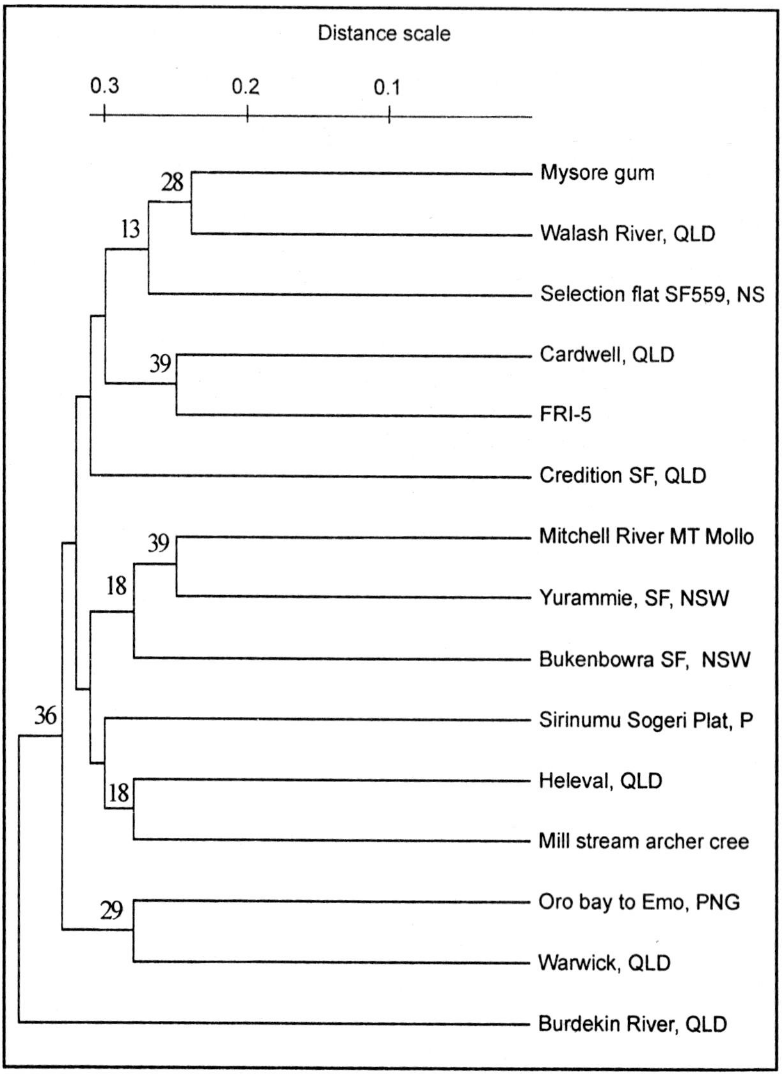

Fig. 2. UPGMA dendrogram showing genetic relationship among fifteen provenances of *E. tereticornis*.

Table 3 : Similarity index of fifteen provenances of *E. tereticornis*.

S.No.	Sources	1	2	3	4	5	6	7	8	9	10	11	12	13	14
1	Oro bay to Emo, PNG	1.00													
2	Sirinumu Sogeri Plat,PNG	0.47	1.00												
3	Warwick,QLD	0.61	0.55	1.00											
4	Yurammie,SF,NSW	0.53	0.6	0.63	1.00										
5	Bukenbowra SF,NSW	0.51	0.51	0.56	0.63	1.00									
6	Selection flat SF559,NSW	0.54	0.51	0.61	0.56	0.59	1.00								
7	Credition SF,QLD	0.51	0.48	0.59	0.59	0.53	0.61								
8	Mitchell River MT Molly,QLD	0.49	0.49	0.61	0.69	0.64	0.58	0.64	1.00						
9	Mill streamarcher creek,QLD	0.48	0.57	0.56	0.63	0.58	0.54	0.56	0.64	1.00					
10	Cardwell,QLD	0.53	0.52	0.6	0.58	0.52	0.59	0.63	0.61	0.58	1.00				
11	Helevale,QLD	0.5	0.55	0.63	0.63	0.56	0.56	0.59	0.63	0.63	0.62	1.00			
12	Walsh River,QLD	0.54	0.53	0.64	0.61	0.57	0.62	0.66	0.62	0.62	0.61	0.67	1.00		
13	Burdekin River,QLD	0.44	0.45	0.57	0.53	0.5	0.52	0.52	0.54	0.52	0.51	0.6	0.54	1.00	
14	FRI-5	0.55	0.5	0.64	0.62	0.63	0.67	0.6	0.67	0.66	0.69	0.61	0.62	0.6	1.00
15	Mysore gum	0.46	0.5	0.58	0.61	0.6	0.66	0.6	0.6	0.56	0.58	0.6	0.7	0.51	0.63

The second cluster grouped another six provenances namely Mitchell River (MT Mollo), Yurammie (NSW), Bukenbowra (NSW), Sirinumu Sogeri Plat (PNG), Helevale, (QLD) and Mill stream archer creek (QLD). Within this cluster, a maximum genetic similarity was observed between Mitchell river; MT Mollo and Yurammie, (NSW) followed by similarity between Helevale, QLD and Mill Stream Archer Creek, QLD. Broadly, the cluster had two sub clusters each of three provenances. The third cluster on the other hand comprised only two closely related provenances; Oro Bayto Emo (PNG) and Warwick (QLD). Burdekin River (Queensland) emerged as an outlier for all the three clusters, making it genetically most distant from the remaining provenances.

DISCUSSION

RAPD markers are preferred and highly reliable over the conventional morphological and biochemical markers to identify genetic variation since they are not influenced by environment and growth stage of experimental plant material (Rajagopal *et al.*, 2000). Standardization of RAPD assay is an essential pre-requisite for obtaining reliable amplification profiles. Our experimentation indicated that both quantity and quality of template DNA played a key role in determining consistency of RAPD profiles. Besides, concentration of Mg ions was also found important in determining the reproducibility of the RAPD profiles. It was found that 5 ng DNA and 3 mM $MgCl_2$ was the best combination for getting consistent and reproducible results. Rajgopal *et al.* (2000) also emphasized importance of DNA quality and Mg ion concentration for reproducibility of RAPD profiles in *Populus ciliata*.

Varying levels of genetic similarity was observed among the 15 provenances of *E. tereticornis*, for the plant material used in the study represents a wide geographic range of longitude and altitude. High genetic variation is attributed to the large geographical distribution and the high cross-pollinating nature of the species. The existence of phenotypic variation in the above *E. tereticornis* provenances has been reported earlier (Ginwal *et al.*, 2005). Using RAPD markers, it is further proved that the above provenances have considerable levels of genetic variation. Establishment of such provenance relationship through DNA based markers has been performed in *Salix* (Barker *et al.*, 1999), Norway spruce (Scheepers *et al.*, 1997), *Mangifera indica* (Schnell *et al.*, 1995) and Himalayan poplar (Rajgopal *et al.*, 2000).

The dendrogram also reveals that no two provenances within the given sources were 100% similar. However, the RAPD data provided evidence for very close resemblance of Mysore gum with Walsh River (QLD) suggesting a strong genetic relationship between the two sources. The "Mysore gum" (*Eucalyptus* hybrid) represents about half of the eucalypts planted in many parts of India (Jacobs, 1981). Initially, the plantations were raised through seeds collected from few *Eucalyptus* hybrid trees located in early introductory trials in Nandi Hills (Mysore, India), the original seed source being Australia (Pryor, 1966; Chaturvedi, 1976). It has also been proposed that "Mysore gum" is mainly *E. tereticornis* but, due to the continuous process of natural selection and occasional natural hybridization with other species viz., *E. robusta* and *E. camaldulensis*, the " Mysore gum" race is known as 'eucalyptus hybrid' (Pryor, 1966). Thus, the close genetic relationship of Mysore gum with the Australian provenances is expected as revealed in the present study.

In the present investigation, the Burdekin river (QLD) provenance did not cluster with any of the sources. This provenance has an inherent capacity to grow fast in comparison to others as revealed by morphological data (Ginwal *et al.* 2005). Based on growth, the north Queensland provenances viz., Walsh River and Burdekin River, performed better (Ginwal *et al.* 2005).

Genetic variation in provenances of *E. tereticornis* revealed by RAPD analysis could possibly due to the fact that this species grows over a wide range of rainfall, temperature and soil types in Australia and Papua New Guinea. Cluster analysis also reflected no clear grouping according to geographic location of the provenances. The above provenances might have experienced marked differences in selective pressure. Results obtained from the above RAPD assay also revealed that there is sufficient genetic variability in the *E. tereticornis* provenances, which can be used for further selection and hybridization work.

ACKNOWLEDGEMENT

The authors are grateful to Dr. Chris Harwood, Australian Seed Centre, CSIRO, Canberra, Australia for providing seed lots of *Eucalyptus tereticornis.*

References

Barker, H.A.J., Matthes, M., Arnold, G.M., Edwards, K.J., Ahman, L., Larsson, S. and Karp, A. (1999) Characterization of genetic diversity in potential biomass willows (Salix spp.) by RAPD and AFLP analyses. *Genome* 42: 173-183.

Castiglione, S., Wang, G., Damini, G., Bandi, C., Bisoffi S. and Sala, F. (1993) RAPD fingerprints for identification and for taxonomic studies of elite poplar (Populus spp.) clones. *Theor. Appl. Genet.* 87: 54-59.

Chaturvedi, A.N. (1976) Eucalyptus in India. *Indian For.* 102: 57-63.

Eldridge, K., Davidson, J,, Harwood, C. and Wyk, G. (1994) *Eucalypt Domestication and Breeding*. Clarendon press. Oxford.

Evans, J. (1992) *Plantation Forestry in the Tropics, 2nd Edition*. Oxford Science Publications, Oxford.

Ginwal, H.S., Kumar, P., Sharma, V.K., Mandal, A.K. and Harwood, C.E. (2005) Genetic variability and early growth performance of *Eucalyptus tereticornis* Sm. in provenance cum progeny trials in India. *Silva. Genet.* 53: 148-153.

Jacobs, M.R. (1981) *Eucalypts for Planting*. FAO Forestry Series No. 11. FAO, Rome.

Keil, M. and Griffin, A.R. (1994) Use of random amplified polymorphic DNA (RAPD) markers in the discrimination and verification of genotypes in *Eucalyptus*. *Theor. Appl. Genet.* 88: 116-122.

Mosseler, A., Egger, K.N. and Hughes, G.A. (1992) Low level of genetic diversity in red pine confirmed by random amplified polymorphic DNA markers. *Can. J. For. Res.* 22: 1132-1137.

Nei, M., and Li, W.H. (1979) Mathematical model for studying genetic variation in terms of restriction endonucleases. *Proc. Nat. Acad. Sci., USA* 76: 5269-5273.

Pryor, L.D. (1966) A report on past performance and some current aspects of the cultivation of quick-growing species (mainly Eucalyptus) in India. *Indian For.* 92: 615-622.

Rajagopal, J., Bashyam, L., Bhatia, S., Khurana, D.K., Srivastava, P.S. and Lakhsmikumaran, M. (2000) Evaluation of genetic diversity in the Himalayan Poplar using RAPD markers. *Silva. Genet.* 49: 60-66.

Scheepers, D., Eloy, M.C. and Briquet, M. (1997) Use of RAPD patterns for clone verification and in studying provenance relationship in Norway spruce (*Picea abies*). *Theor. Appl. Genet.* 94: 480-485.

Schnell, R.J., Ronning, C.M. and Knight Jr., R.J. (1995) Identification of cultivars and validation of genetic relationship in *Mangifera indica* L. using RAPD markers. *Theor. Appl. Genet.* 90: 269-274.

Stange, C., Prehn, D. and Johnson, P.A. (1998) Isolation of *Pinus radiata* genomic DNA suitable for RAPD analysis. *Plant Mol. Biol. Rep.* 16: 1-8.

Whitkus, R., Cruz, M., Mota Bravo, L. and Gomez-Pompa, A. (1998) Genetic diversity and relationships of cacao (*Theobroma cacao* L.) in southern Mexico. *Theor. Appl. Genet.* 96: 621-627.

Williams, J.G.K., Kubelik, A., Livak, K.J., Rafalski, J.A.and Tingey, S.V. (1990) DNA polymorphisms amplified by arbitrary primers are useful as genetic markers. *Nucleic Acids Res.* 18: 6531-6535.

Zobel, B.J., Van Wyk, G. and Stahi, P. (1987) *Growing Exotic Forests.* A Wiley Inter Science Publication. John Wiley and Sons, NY, USA.

Forest Biotechnology in India © 2008
Editors: S.A. Ansari, C. Narayanan & A.K. Mandal, pp.89-96
Satish Serial Publishing House, Delhi

9

Studies on Identification and Genetic Relationships of Six *Albizia* Spp. Using RAPD Markers

SUBHASHREE APARAJITA AND G.R.ROUT

Plant Biotechnology Division, Regional Plant Resource Centre, Department of Forest & Environment
Bhubaneswar - 751 015, Orissa, India

ABSTRACT

Studies were undertaken for identification and assessment of genetic relationships in six species of *Albizia*, viz., A. *procera*, *A. lebbek*, *A. amara*, *A. stipulata*, *A. thomsonii* and *A. odoratissima*, using the random amplified polymorphic DNA (RAPD) marker. RAPD analysis with 50 decamer primers distinguished all the six species. Based on their polymorphic nature, nine RAPD primers were further selected and used for evaluation of genetic variability among individuals of populations of the six *Albizia* spp. A total of 110 distinct DNA fragments (bands), ranging from 0.3 - 3.0b was amplified. Analysis of genetic similarity on the basis of presence or absence of bands revealed a narrow range of variability within each species. UPGMA cluster analysis showed two major clusters. *A. lebbek* clustered separately from the other five species. The cluster of the remaining five species divided into two minor clusters with *A. thomsonii* and A. *procera* in one group and A. *amara*, *A. odoratissima* and *A. stipulata* in the other. Thus, RAPD markers have the potential for identification and assessment of genetic similarity in *Albizia* spp. The above information could be useful for genetic improvement and conservation of the above six *Albizia* spp.

Keywords: *Albizia*, genetic similarity, identification, RAPD marker.

INTRODUCTION

The genus *Albizzia*, belonging to family Leguminosae, is widely distributed in the tropical and sub-tropical regions of Asia, Africa and Australia. Several species of *Albizia* yield high quality timber and some species have medicinal value. Plant extracts of some *Albizzia* spp. have antiseptic, anti-dysenteric and anti-tuberculosis properties. The seeds cure piles, diarrhea and gonorrhea (Anonymous, 1989). Germplasm characterization is an important prerequisite for conservation and utilization of plant genetic resources. Conventionally, morphological characters like habit, leaf type, floral morphology and fruit characters have been used to define genetic variation. Molecular techniques help researchers not only to identify genotypes, but also to assess and exploit genetic variability (Whitkus *et al.*, 1994). DNA fingerprinting of the genetic resources of the timber-yielding plants, is a necessity for generation of a molecular database that catalogues and allows utilization of information in a systematic manner. Moreover, the commercial value associated with identification of useful traits creates a direct value for gene banks, ensuring long-term preservation of a collection. The genetic relationship can be useful for designing strategies for gene introgression and breeding programs to produce desired recombinant hybrid genotypes with both high timber quality and high biomass yields. Management decisions for conservation of taxa largely depend on better understanding of their biology and genetic variability that influence their survival. Conservation management decisions for rare taxa, however, often have to be made quickly without adequate ecological data or information on genetic variability (Gaston and Kunin, 1997).

Random amplified polymorphic DNA (RAPD) analysis has been proved useful for estimating genetic diversity particularly to assist in the conservation of rare species and plant genetic resources (Andersen and Fairbanks, 1990). RAPD analysis, in particular, has proven to be a rapid and efficient means of genome mapping (Williams *et al.*, 1990) and is well suited for genetic resource characterization (Andersen and Fairbanks, 1990). DNA fingerprinting generated by the polymerase chain reaction (PCR) using arbitrary primers has provided a useful tool for the detection of DNA polymorphism in a number of tree species (Carlson *et al.*, 1991; Byrne *et al.*, 1995; Verhaegen and Plomion, 1996). In this context, information on genetic relationships and diversity of available germplasm is essential for characterization of germplasm and optimization of hybridization and selection. In this study, RAPD markers were used to evaluate the extent of genetic variation among six species of *Albizzia.*

MATERIALS AND METHODS

Plant material

Leaf material was collected from 15 individuals each of *A. procera, A. lebbek, A. amara, A. stipulata, A. odoratissima* and *A. thomsonii* from a natural stand at Chandaka Reserve Forest, Bhubaneswar, and Regional Plant Resource Centre, Bhubaneswar. The morphological characteristics of the above *Albizia* species are presented in Table 1.

Table 1: Morphological characteristics of six species of *Albizzia.*

Plant Characters	*A. amara*	*A. odoratissima*	*A. stipulata*	*A. procera*	*A thomsonii*	*A. lebbek*
Habit	Deciduous	Deciduous	Deciduous	Deciduous	Deciduous	Deciduous
Plant height (m)	12	25	20	20	20	25
Type of bark	Smooth, brown	Rough, greyish to black	Dark grey or brown	Whitish	Light colour and tawny	Rough, brownish gray
Leaf length (cm)	12	15	15 - 25	14	10 - 15	14 – 15
No. of pinnae (pairs) / leaf	9 - 13	6 - 16	6 - 13	3 – 6	3 - 10	4 - 8
Pinnae Length (cm)	1.5 – 5	6 – 13	4.5 – 10	12 – 24	5 – 15	11 – 19
Leaflet no. (pairs) / pinnae.	20 - 25	20 -25	25 - 45	12 – 20	10 - 25	10 - 16
Leaf shape	Narrow, Elliptical	Elliptical	Obliquely oblong	Rhomboid oblong	Oblong	Obtuse
Leaflet length (mm)	8	1.5 - 2	6 - 12	2 – 5	1.5 - 2	1.5 - 3
Presence of Gland	Petiole having gland above the middle	Petiole having gland below the middle	-	Below the upper 1 - 3 pairs of leaflet	-	-
Inflorescence	Racemes	Umbellate	Head	Spherical head	Head	Umbellate
Inflorescence length (cm)	2 - 3	2.0	2 – 3	1 - 1.5	2 - 2.5	2.5 - 3

Contd...

Table 1. Contd...

Petal colour	Cream	Cream	Greenish white	White	Greenish yellow	Greenish White yellow
Pod shape	Flat, Compressed, strongly nerved	Flat, Compressed, Strongly nerved	Flat, Nerved	Flat	Flat , Beaked	Thin, Flat
Pod length (cm)	12 - 20	15 - 25	8 - 10	10 - 12	8 - 10	15 – 25
Seed shape	Ovoid	Ovoid	Ovoid	Ovoid	Ovoid	Ovate
No. of seeds / Pod	10	10 - 12	8 - 10	6 – 8	6 - 8	6 – 8
Seed size (cm)	1 - 1.2	0.8 –1.1	0.5 - 0.7	0.5 - 0.8	0.6 – 0.8	0.8 – 1.0

DNA isolation

DNA was extracted from young leaves using the method described by Doyle and Doyle (1990) with modifications. Two grams of fresh leaf material were washed in distilled water and subsequently sterilized with 80% (v/v) ethanol and then ground in liquid nitrogen. Ten ml of preheated extraction buffer (2% CTAB (w/v), 0.2% b-mercaptoethanol (v/v), 100 mM Tris-HCl pH 8.0, 2 mM EDTA, 1.4 M NaCl) was then added. After removal of protein and other contaminants, the DNA pellet was finally resuspended in 200 to 300 µl of Tris-EDTA. DNA quantification was done by visualization under UV light, after electrophoresis on 0.8% agarose gels.

Primer screening

Forty 10-mer primers, corresponding to kits A, B, D, and N from Operon Technologies (Alameda, California) were initially screened to determine the suitability of each primer. Primers were selected for further analysis based on their ability to detect distinct, clearly resolved and polymorphic amplified products among the six species of *Albizzia*.

PCR amplification

Polymerase chain reactions (PCR) with each RAPD primer were carried out in a final volume of 25 µl containing 20 ng template DNA,

100 mM of dNTPs, 20 ng of primer (Operon Technology, USA), 1.5 mM $MgCl_2$, 1x Taq buffer and 0.5 U Taq DNA polymerase (M/S Bangalore Genei, India). DNA amplification was done in a PTC 100 thermal cycler (MJ Research, USA) programmed for a preliminary 4 min denaturation step at 94°C, followed by 45 cycles of denaturation at 94°C for 1 min, annealing at 37°C for 1 min and extension at 72°C for 2 min, finally at 72° C for 10 min. Amplification products were separated alongside a molecular weight marker (100 bp ladder, MBI Fermentas, USA) by electrophoresis on 1.2% agarose gels run in 1x TAE buffer, stained with ethidium bromide and visualized under UV light. Gel photographs were scanned through Gel Doc System (Gel Doc. 2000, BioRad, USA) and the amplification products were analyzed using the software Quantity-one (BioRad, USA).

Data analysis

Each experiment was conducted thrice. Data were recorded as presence (1) or absence (0) of band products from the examination of photographic negatives. Each amplification fragment was named by the source of the primer, the kit letter or number, the primer number and its approximate size in base pairs. Bands with similar mobility to those detected in the negative control, if any, were not scored. Similarity index was estimated using the Dice coefficient of similarity (Nei and Li, 1979). The similarity matrices were used to generate a dendrogram by UPGMA (Unweighted Pair-Group Method Arithmetic Average) method using NTSYS-PC, version 1.80 (Rohlf, 1995).

RESULTS

The primer screening step resulted in nine 10-mer primers which showed polymorphisms within the six species of *Albizzia*. Twenty primers showed polymorphism, but could not distinguish among species; five primers showed monomorphic bands among the species and ten gave no amplification products. The reproducibility of DNA amplification was tested through three independent PCR experiments. Only those bands that were consistently amplified were considered for further analysis. Bands with same mobility were considered as identical fragments, regardless of their staining intensity. The results indicated that nine RAPD primers were able to detect polymorphism among all the species of *Albizzia*. The primers OPA-04, OPD-18 and OPD-20 produced maximum numbers of bands (Table 2). A total of 110 amplified fragments were scored across six species for the nine selected primers. From the results of the RAPD profiling, it was observed that some species showed high numbers of amplified bands with the primers OPA-04, OPD-18, OPD-20 and OPN-16 while others showed fewer numbers of amplified bands with the same primers. Nine 10-mer primers also generated unique patterns, differentiating all six species from each other. The similarity matrix obtained after multivariant analysis using Nei and Li' s coefficient is presented in Table 3. The cluster analysis indicates that the six species of *Albizia* formed two major clusters. The first cluster consisted of only one species, *A. lebbek.* The second major cluster included five species, *A. odoratissima, A. amara, A. stipulata. A. procera* and *A. thomsonii.* This cluster consisted of two sub-clusters i.e., *A.procera* and *A.thomsonii* in one sub-cluster and *A. stipulata, A.amara* and *A.odoratissima* in another sub-cluster.

Table 2: Total number of amplified fragments and number of polymorphic fragments generated by PCR using selected random primers in six *Albizia* species.

Name of primer	Sequence	Total No. amplification products	No. of polymorphic products	Size range (kb)
OPA 02	5′-TGCCGAGCTG-3′	11	11	0.30 - 2.0
OPA 04	5′-AATCGGGCTG-3′	15	15	0.40 - 3.0
OPA 13	5′-CAGCACCCAC-3′	10	10	0.65 - 2.7
OPA 18	5′-AGGTGACCGT-3	9	09	0.3 - 1.5
OPD 05	5′-TGAGCGGACA-3′	11	11	0.55 - 2.4
OPD 18	5′-GAGAGCCAAC-3′	15	15	0.35 - 2.5
OPD 20	5′-ACCCGGTCAC-3′	15	15	0.35 - 2.5
OPN 06	5′-GAGACGCACA-3′	13	13	0.40 - 3.0
OPN 16	5′-AAGCGACCTG-3′	12	11	0.30 - 2.2

Table 3 : Similarity matrix for Nei and Li's coefficient of six species of *Albizzia.*

	A1	A2	A3	A4	A5	A6
A1	1.0					
A2	0.52	1.0				
A3	0.63	0.40	1.0			
A4	0.50	0.47	0.58	1.0		
A5	0.51	0.50	0.61	0.50	1.0	
A6	0.61	0.52	0.47	0.56	0.51	1.0

Note: A1 – *Albizia amara* , A2 – *Albizia lebbek,* A3 – *Albizia odoratissima,* A4 – *Albizia procera,* A5 – *Albizia stipulata,* A6 – *Albizia thomsonii*

DISCUSSION

The present investigations showed that it is possible to identify and assess genetic relationship among six species of *Albizzia*. After screening several Operon RAPD primers, nine informative primers were found capable of detecting DNA polymorphism among six species of *Albizzia*. Further analysis showed that RAPD profiles clearly differentiated all six species of *Albizzia*. RAPD markers have been used for identification of different plant species (Waycott, 1998; Rout *et al.*, 2003; Belaj *et al.*, 2004). Further, the taxa-specific bands can be utilized for defining uniqueness, which will be helpful in species and taxa identification. In our present study, unique patterns were observed, differentiating *Albizia* species which can possibly be used in programs related to conservation of germplasm, breeding and management of genetic resources of *Albizia* spp.

Thus, RAPD markers provided valuable information for identification of species and assessment of genetic relatedness within species for conservation of *Albizia* spp. Further studies on understanding the population genetic structure of the different species would provide an important input for developing appropriate management strategies.

ACKNOWLEDGEMENT

The authors wish to acknowledge the help of the Department of Biotechnology, Ministry of Science and Technology, Government of India for financial assistance under grant No. BT/PR/3750/AGR/16/309/2003-04.

References

Andersen, W.R. and Fairbanks, D.J. (1990) Molecular markers: Important tools for plant genetic resource characterization. *Diversity* 6: 51-53.

Anonymous. (1989) *The Wealth of India: Raw Materials, Vol. A-B*. Council of Scientific Industrial Research, New Delhi.

Belaj, A., Satovic, Z., Rallo, L. and Trujillo, I. (2004) Optimal use of RAPD markers for identifying varieties in olive (*Olea europaea* L.) germplasm collections. *J. Am. Soc. Hort. Sci.* 129: 266-270.

Bretting, P.K. and Widrelechner, M.P. (1995) Genetic markers and horticultural germplasm management. Hort. Sci. 30: 1349-1356.

Byrne, M., Murrell, J.C., Allen, B. and Moran, G.F. (1995) An integrated genetic linkage map for eucalyptus using RFLP, RAPD and isozyme markers. *Theor. Appl. Genet.* 91: 869-875.

Carlson, J.E., Tulsieram, L.K., Glaubitz, J.C., Luk, V.W.K., Kauffeld, C. and Rutledge, R. (1991) Segregation of random amplified DNA markers in F1 progeny of conifers. *Theor. Appl. Genet.* 83: 194-200.

Doyle, J.J. and Doyle, J.L. (1990) Isolation of plant DNA from fresh tissue. *Focus* 12: 13-15.

Gaston, K.J. and Kunin, W.E. (1997) Concluding comments. In: *The Biology of Rarity* (eds. Kunin, W.E., Gaston, K.J.) Chapman and Hall, London, pp. 262- 272.

Nei, M. and Li, W.H. (1979) Mathematical modes for studying genetic variation in terms of restriction endonucleases. *Proc. Natl. Acad. Sci.,* USA 76: 5269 – 5273.

Rohlf, F.J. (1995). NTSYS-pc Numerical Taxonomy and Multivariate analysis system. Version 1.80, Exter Software, Setauket, New York.

Rout, G.R., Bhattacharya, D., Nanda, R.M., Nayak, S. and Das, P. (2003) Evaluation of genetic relationships in *Dalbergia* species using RAPD markers. *Biodiv. Conserv.* 12: 197-206.

Sneath, P.H.A. and Sokal, R.R. (1973) *Numerical Taxonomy. The Principles and Practice of Numerical Classification.* W.H.Freeman & Co., San Francisco, California.

Verhaegen, D. and Plomion, C. (1996) Genetic mapping in *Eucalyptus urophylla* and *Eucalyptus grandis* using RAPD markers. *Genome* 39: 1051-1061.

Virk, P.S., Ford-Lloyd, B.V., Jackson, M.T. and Newbury, H.J. (1995) Use of RAPD for the study of diversity within plant germplasm collections. *Heredity* 74: 170-179.

Waycott, M. (1998) Genetic variation, its assessment and implications to the conservation of sea grassess. *Mol. Ecol.* 7: 793 – 800.

Whitkus, R., Doebley, J. and Wendel, J.F. (1994) Nuclear DNA Markers in Systematics and Evolution. In: DNA-based markers in plants (eds. Phillips, L. and Vasil, I.K.) Kluwer Academic Publishers, Amsterdam, pp. 116-141.

Williams, J.G.K., Kubelik, A., Livak, K.J., Rafalski, J.A.and Tingey, S.V. (1990) DNA polymorphisms amplified by arbitrary primers are useful as genetic markers. *Nucleic Acids Res.* 18: 6531- 6535.

❑❑❑

Forest Biotechnology in India © 2008
Editors: S.A. Ansari, C. Narayanan & A.K. Mandal, pp.97-107
Satish Serial Publishing House, Delhi

10

Allelic Diversity in Cellulose Synthase Gene in *Eucalyptus Tereticornis*: Implications in Association Mapping

M. GHOSH, P. CHEZHIAN, R. KAMALAKANNAN,
R. YASODHA, M. VARGHESE AND K. GURUMURTHI

Division of Plant Biotechnology, Institute of Forest Genetics and Tree Breeding, Coimbatore 641 002, India

ABSTRACT

Eight international provenances of *Eucalyptus tereticornis* were analyzed for the allelic diversity existing in the cellulose synthase genes. Three primers were developed targeting the SSR flanking regions in cellulose synthase genes of *Hordeum vulgare, Populus tremuloides* and *Gossypium hirsutum* and designated as *CesA1, CesA2* and *CesA3* with expected product size at 241 bp, 139 bp and 153 bp, respectively. Allelic diversity was observed within and between the eight provenances using *CesA2* with amplicons ranging from 170 bp to 300 bp with several non-specific bands. Specific bands for *CesA2* at 174 bp and 176 bp were observed in all the three individuals of Orobay provenance and two individuals of Cardwell and North Kennedy River provenance. Allelic diversity was also observed within the three individuals of S.W. of Mt. Garnet provenance. The results were further corroborated in all the five families of Orobay provenance consisting of four individuals each and the 174 bp and 176 bp products amplified in almost all the individuals. Allelic diversity within and between almost families of Orobay was also observed. The cellulose synthase gene diversity was correlated with cellulose content in the wood samples.

Keywords : Allelic diversity, cellulose synthase gene, *Eucalyptus*, provenance, genetic diversity.

INTRODUCTION

The major goal of modern biology has been to understand the genetic basis of naturally occurring variation in complex traits. Linkage disequilibrium (LD) or "association mapping" of candidate gene associations is an emerging approach for identifying the genes underlying such phenotypic variations (Ardlie *et al.*, 2002). In contrast to the more established approach of quantitative trait locus (QTL) mapping, this technique infers associations between genotype and phenotype variation resulting from recombination. While QTL mapping considers only variation between two crossed individuals and relies solely on recombination events observed in their progeny. LD mapping exploits the phenotypic and genetic variation present across a natural population and draws inferences on the basis of past recombination events (Nordborg and Tavare, 2002; Borevitz and Nordborg, 2003). This method has been successfully applied in studies of *Drosophila* (Long *et al.*, 1998), maize (Thornsberry *et al.*, 2001; Wilson *et al.*, 2004), wheat (Breseghello and Sorrells, 2005) and *Arabidopsis* (Olsen *et al.*, 2004) to identify genes and specific polymorphisms within genes that underlie natural phenotypic variation. The only study on "association mapping" on candidate genes in forest tree species is in *Eucalyptus* spp. wherein allelic variant of the CCR gene from the phenyl propanoid pathway was demonstrated to have significant association with microfibril angle (MFA) in a natural population of *E. nitens*. The association was further confirmed in-full sib families of *E. globulus* and. *E. nitens* (Thumma *et al.*, 2005). In the present study a similar approach was followed and allelic diversity of cellulose synthase gene, the major gene in the cellulose biosynthetic pathway was analyzed in a natural unstructured population of *E. tereticornis* and the genotypic data was correlated with the cellulose content of the wood samples. This is the first attempt targeting the association of cellulose synthase allelic variation with the phenotype (cellulose content in wood) in *E. tereticornis.*

MATERIAL AND METHODS

Plant material

In the present study, eight international provenances of *E. tereticornis* (with seed source from CSIRO, Australia) presently in a trial at Karunya (Coimbatore, Tamil Nadu) were used for population analysis (Table 1).

Table 1: *Eucalyptus tereticornis* provenances used in the study

Seedlot No.	Provenance	Origin*
13277	Cardwell	QLD
13399	Orobay	PNG
13418	Sogeri Plateu	PNG
13847	Palmer R	QLD
16547	Normanby R	QLD
17864	N Kennedy R	QLD
16554	SW of Mt. Garnet	QLD

* QLD - Queensland Australia PNG - Papua New Guinea

Identification of unstructured population of E. tereticornis

The prerequisite for "association mapping" is a population with no significant structure. Hence to determine the suitability of the above-mentioned population for association mapping ten individuals from each provenances were profiled with ISSR primers and analyzed for their structure. Genomic DNA was extracted from young leaf tips as described by Doyle and Doyle (1987) with minor modifications from ten individual representatives of each provenance.

ISSR-PCR electrophoresis

DNA from all individuals were amplified using the ISSR primers detailed in Table 1. The PCR mixture (10 μl) contained 1.0 μl 10x PCR buffer, 2.5 mM $MgCl_2$, 40 μM dNTP mix, 0.3 U Taq DNA polymerase (Bangalore Genei, India), 100 nM ISSR primer (Sigma-Aldrich, USA) and 15 ng of template DNA. A total of four primers were used in the present study. ISSR amplifications were carried out in a programmable thermal cycler (PTC-200, MJ Research) in three steps: Step 1 was 3 min at 94°C. Step 2 was 35 cycles of 30 seconds at 94°C for denaturation, 30 seconds at 50°C for annealing, and 1 min at 72°C for polymerization with final step of extension at 72°C for 10 min.

Fragments were separated in 2 % agarose gel with 1x TAE buffer at 70V electric field for two hours along with 1 kb ladder (Gibco BRL) for molecular weight determination. The gel profiles were viewed under UV-transilluminator and documented using Kodak-DC290 digital camera.

Table 2: ISSR primers selected for diversity analysis

Primer code	Nucleotide sequence
5' anchored	
$R(CA)_7$	5' GRTRCYGRTRCACACACACACACA 3'
$T(GT)_9$	5' CRTAYGTGTGTGTGTGTGTGT 3'
$TA(CAG)_4$	5' ARRTYCAGCAGCAGCAG 3'
$RA(GCT)_6$	5' AYARAGCTGCTGCTGCTGCTGCT 3'
3' anchored	
$(GA)_8R$	5' GAGAGAGAGAGAGAGARGY 3'
UBC810	5' GAGAGAGAGAGAGAGAT 3'
UBC842	5' GAGAGAGAGAGAGAGAYG 3'

Data analysis

ISSR fragments were scored as presence (1) or absence (0) in each individual and a similarity matrix was derived from the binary data using Nei and Li's similarity indices (Nei and Li, 1979). The similarity matrix was used in a UPGMA cluster analysis and a principal coordinate analysis, by employing the software NTSYS-pc (Rohlf, 1992). Binary data were used to calculate the genetic identity/diversity (Nei, 1972) among the eight provenances using POPGENE, Version 1.32 (Yeh *et al.*, 1997), assuming Hardy-Weinberg equilibrium.

Genotyping of E. tereticornis provenances with cellulose synthase primers

Genomic DNA was extracted from young leaf tips of three individuals belonging to eight provenances and nineteen individuals of Orobay provenance according to Doyle and Doyle (1987) with minor modifications. Template DNA (50 ng) was amplified in a total volume of 10 μL containing 1.0 μL 10x PCR buffer (Bangalore Genei Ltd., India), 40 nM of each primer (forward and reverse primer), 0.4 mM dNTP mix, and 0.3 U of Taq DNA polymerase (Bangalore Genei Ltd., India). Since no sequence data was available for cellulose synthase genes from eucalypt species, primers were designed for SSRs in cellulose synthase genes from *Populus tremuloides, Hordeum vulgare* and *Gossypium hirsutum* (Table 3). PCR was performed in a programmable thermal cycler (PTC-200; MJ Research). The conditions for amplifications were as follows: Initial denaturation (5 min, 94°C), followed by 30 cycles consisting of denaturation (1 min, 94°C), annealing (1 min; temperature conditions as given in Table 3), extension (2 min, 72°C) and a final extension (10 min, 72°C). The amplified fragments were separated in a 6% denaturing polyacrylamide gels and stained with silver nitrate (Bassam et al., 1991). The gel profiles were documented with Camedia digital camara.

Table 3: Specific markers for Cellulose Synthase genes

Gene	Accession Number	Primer Sequence (5 - 3)	Expected length (bp)	Annealing temperature (°C)
CesA1 (Hordeum vulgar)	AY483152	5' GAAGGTGTTTTTGTGCCCAT 3' 5' ACTTGCTGGCTGGCTGTATT 3'	241	45
CesA2 (Populus tremuloides)	AF417485	5' GACATGCAACTGCTTGCCTA 3' 5' TACCCTCCAAAGTCCCACAG 3'	139	54
CesA3 (Gossypium hirsutum)	AF254895	5' GGTTTATTGGTGTGTTGGGG 3 5' TCTCAATTTTGAATAAGGACCAAA 3'	153	45

Estimation of cellulose content in wood samples

The eight *E. tereticornis* provenances (Table 1) were analysed for cellulose content using wood sample from three individual each and from nineteen individuals of five families belonging to Orobay provenance. Wood core samples were taken from three individuals of a given provenance at a height of 1.37 m. Total cellulose content in the oven dried wood samples was estimated according to the protocol described by Updegroff (1969). 200 mg of wood samples were ground and hydrolyzed in 3 ml of acetic/nitric reagent (150 mL/15 mL) and incubated at 100°C for 30 min. The extracts were cooled, centrifuged and the resulting pellet were washed with distilled water and subsequently dissolved in 10 ml of 67 per cent sulfuric acid. The samples were diluted 100 times and from the resulting solution one ml fractions were analyzed for cellulose content (Table 4) by adding anthrone reagent. The color intensity was measured at 630 nm wavelength and cellulose content was determined using a standard graph.

Table 4: Cellulose content in wood samples of different provenances of *E. tereticornis.*

Provenance	Sample Number	Cellulose Content (mg / 100mg)
Palmer River	1	1.28
	2	1.17
	3	0.79
Sogeri Plateau	1	4.54
	2	7.13
	3	6.29
Orobay	1	7.25
	2	8.25
	3	8.75
Cardwell	1	3.79
	2	3.25
	3	4.83
IFGTB	1	5.00
	2	4.25
	3	4.25
Normanby River	1	2.42
	2	2.25
	3	2.25
N Kennedy River	1	4.46
	2	5.13
	3	4.88
SW Mt. Garnet	1	12.75
	2	4.96
	3	1.67

Correlation analysis of genotypic data with cellulose content

The correlation of the genotypic data with the cellulose content was done using the SPSS ver. 11.0 software.

RESULTS

Identification of unstructured population

Analysis of genetic relationship within and between eight *E. tereticornis* provenances showed significant polymorphism with four ISSR primers. A total of 92 loci were amplified from 82 individuals belonging to eight provenances and PCR products varied in length between 463 bp and 1579 bp. The total number of amplified loci per primer for all eight provenances varied from 119 (with UBC842) to 167 (with TA $(CAG)_4$). Minimum percent polymorphism (89.2) was observed in Sogeri Plateu provenance while maximum percent polymorphism (100) was observed in Normanby River and Palmer River provenances.

UPGMA dendrogram (Fig.1a) separated the eight provenances into two main clusters as PNG (Papua New Guinea, Australia) and QLD (Queensland, Australia) based on geographical location. The QLD provenances cluster further divided into three sub clusters and among the QLD provenances, N Kennedy River and Normanby River showed maximum genetic distances. The PCO plot revealed the unstructured nature of the population (Figure 1 b).

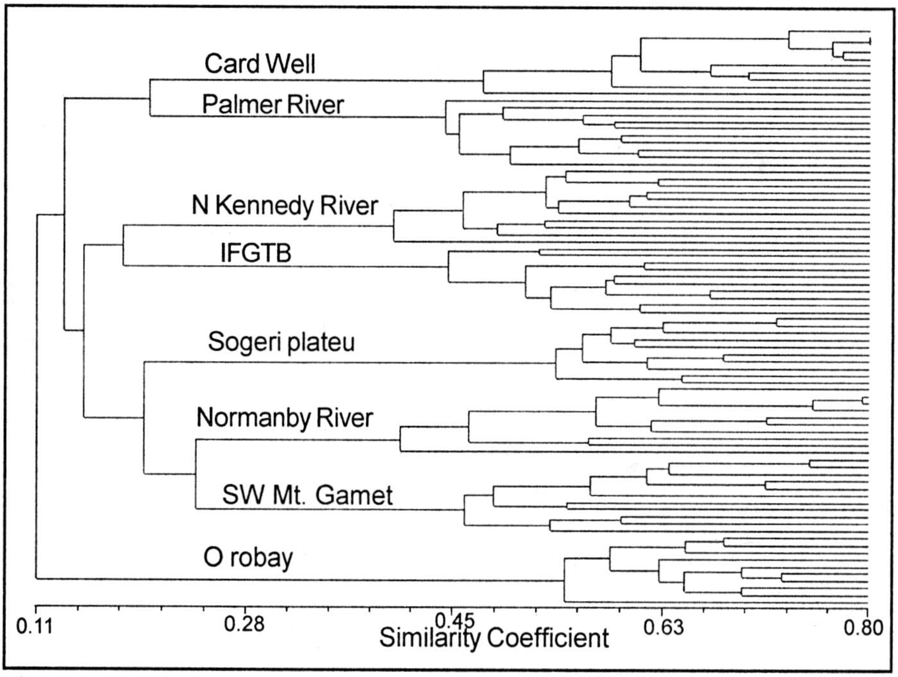

Fig. 1a. Dendrogram showing the similarity among eight provenances of *E. tereticornis*

Genotyping of *E. tereticornis* provenances with cellulose synthase primers and correlation analysis

CesA1 amplified 8 alleles ranging from 150 to 300 bp in all individuals of eight provenances while *CesA2* and *3* amplified 6 and 5 alleles, respectively in the expected range. Specific bands for *CesA2* at 174 bp and 176 bp were observed in all the three individuals of Orobay provenance and two individuals of Cardwell and North Kennedy River provenance. Allelic diversity was also observed within the three individuals of S.W. of Mt. Garnet provenance (Fig. 2 a). The results further corroborated all the five families of Orobay provenance consisting of four individuals each and the 174 bp and 176 bp products amplified in all the individuals except few. Allelic diversity within and between the families of Orobay was also observed (Fig. 2b). The two alleles at 176 bp and 174 bp significantly correlated with high cellulose content using Pearson's correlation (Table 5).

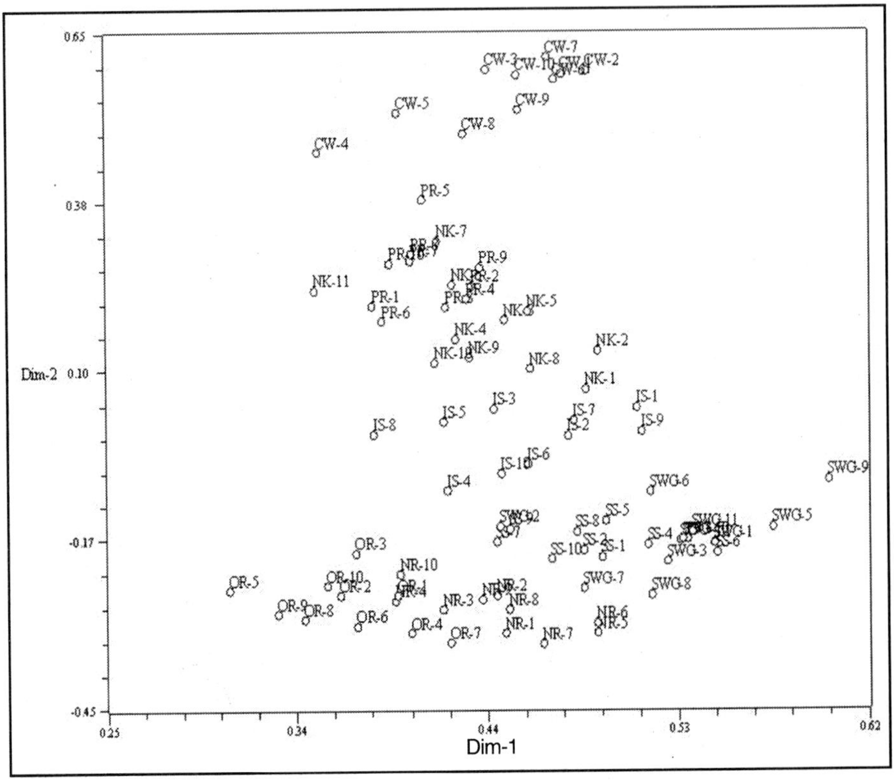

Fig. 1b. PCO plot showing distribution of *E. tereticornis provenances*

Table 5: Correlation (Pearson) analysis of cellulose content and marker loci

	174 bp	176 bp	Cellulose %
174 bp	1.000	0.632**	0.524*
176 bp	0.632*	1.000	0.699**
Cellulose %	0.524*	0.699**	1.000

** Correlation is significant at 0.01 level   * Correlation is significant at 0.05 level

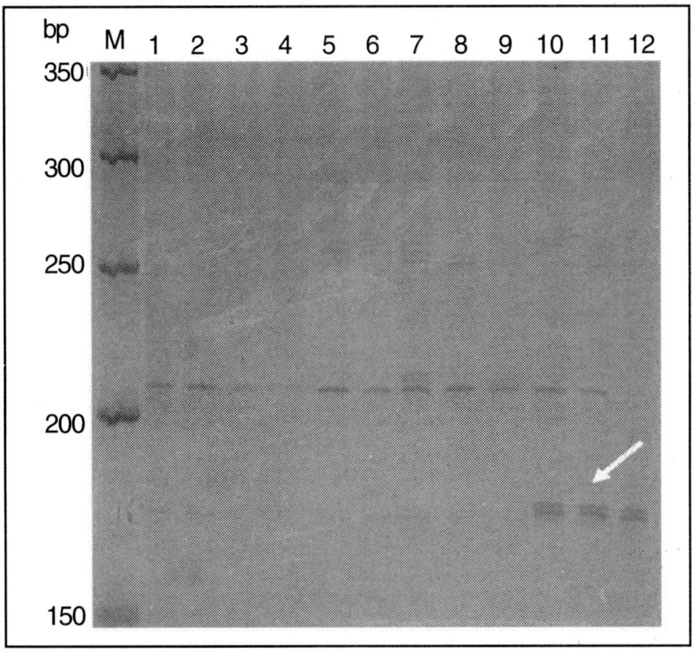

Fig. 2a Amplification of cellulose synthase alleles in *E. tereticornis* provenances using primer *CesA2*

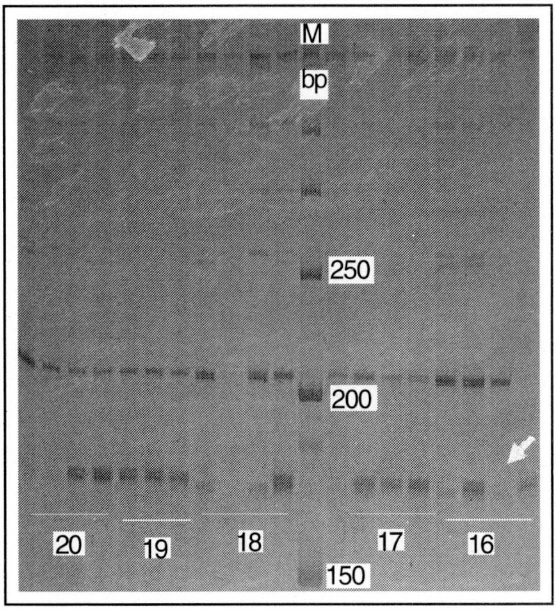

Fig. 2b Amplification of cellulose synthase genes using primer *CesA2* in nineteen individuals of Orobay provenance of *E. tereticornis*

DISCUSSION

The success of association mapping in localizing candidate polymorphisms depends on genomic patterns of linkage disequilibrium in a species and the resulting haplotype structure. Earlier studies on association mapping has been limited to few plant species like maize where six candidate genes involved in kernel composition trait viz., amylase *extender1 ae1, brittle endosperm2 bt2, shrunken1 sh1, sh2, sugary* 1 and *waxy 1* were associated with traits such as protein, oil, starch concentration, pasting properties and amylase levels (Wilson *et al.*, 2004). In wheat, an association of 62 SSR loci on chromosomes 2D, 5A and 5B with kernel morphology and milling quality was analyzed (Breseghello and Sorrells, 2005). In *Arabidopsis*, the *CRY2* allele was found to be significantly associated with early flowering (Olsen *et al.*, 2004) and genes associated with the vernalization response locus *FRI* and the three pathogen resistance loci, *Rpm1, Rps5* and *Rps2* showed significant association with corresponding phenotypes (Aranzana *et al.*, 2005). The only study on Association mapping on candidate genes in forest tree species is in *Eucalyptus* spp. where in allelic variant of the CCR gene from the phenyl propanoid pathway was demonstrated to have significant association with microfibril angle (MFA) in a natural population of *E. nitens*. The association was further confirmed in full-sib families of *E. globulus* and. *E. nitens* (Thumma *et al.*, 2005). In the present study, an attempt was made to study the allelic diversity existing in the natural unstructured population of *E. tereticornis* for cellulose synthase genes and the genotypic data was correlated with cellulose content in the wood. This study forms the basis for association mapping in *E. tereticornis* and also highlights the use of cellulose synthase as a suitable candidate gene.

ACKNOWLEDGEMENT

The authors are grateful to Department of Biotechnology, Government of India for funding the research work.

References

Aranzana, M.J., Kim, S., Zhao, K., Bakker, E., Horton, M. (2005) Genome - wide association mapping in Arabidopsis identifies previously known flowering time and pathogen resistance genes. *PLos Genet.* 1(5): e60.

Ardlie, K.G., Kruglyar, L and Seielstad, M. (2002) Patterns of linkage disequilibrium in the human genome. *Nat. Rev. Genet.* 3: 299- 309.

Bassam, B.J., Caetano-Anolles, G. and Gresshoff, P.M. (1991) Fast and sensivity silver staining of DNA in polyacrylamide gels. *Ann. Biochem.* 196: 80-83.

Borevitz, J.O. and Nordborg, M. (2003) The impact of genomics on the study of natural variation in Arabidopsis. *Plant Physiol.* 132: 718 - 725.

Bressghello, F. and Sorrells, M.E. (2005) Association mapping of kernel size and milling quality in wheat (*Triticum aestivum* L.) cultivars. *Genetics* 172: 1165-1177.

Doyle, J.J. and Doyle, J.L. (1987) A rapid DNA isolation procedure for small quantities of fresh leaf. tissue. *Phytochem. Bull.* 19: 11-15.

Long, A.D., Lyman, R.F., Langley, C.H. and Mackay, T.F.C. (1998) Two sites in the *Delta* gene region contribute to naturally occurring variation in bristle number in *Drosophila melanogaster. Genetics* 149: 999 -1017.

Nei, M. (1972). Genetic distance between populations. *Am. Natur.* 106: 283-292.

Nei, M and Li, W. H. (1979) Mathematical models for studying genetic variation in terms of restriction endonucleases. *Proc. Nat. Acad. Sci. USA.* 76: 5269-5273.

Nordborg, M. and Tavare, S. (2002) Linkage disequilibrium: what history has to tell us. *Trends Genet.* 18: 83 - 90.

Olsen, K.M., Halldorsdottir, S.S., Stinchcombe, J.R., Weinig, C., Schmitt, J. and Purugganan, M.D. (2004) Linkage disequilibrium mapping of Arabidopsis CRY2 flowering time alleles. *Genetics* 167: 1361 - 1369.

Rohlf, F.J. (1992) NTSYS-pc. Numerical taxonomy and multivariate analysis system, Version 1.80.

Thornsberry, J.M., Goodman, M.M., Doebley, J., Kresovich, S., Nielsen, D. (2001) *Dwarf8* polymorphisms associate with variation in flowering time. *Nat. Genet.* 28: 286- 289.

Thumma, B.R., Nolan, M.F., Evand, R. and Moran, G.F. (2005) Polymorphism in *Cinnamoyl COA Reductase (CCR)* are associated with variation in microfibril angle in *Eucalyptus* spp. *Genetics* 171: 1257-1265.

Updegroff, D.M. (1969) Micro determination of cellulose in biological materials. *Annals. Biochem.* 32: 420-424.

Wilson, L.M., Whitt, S.R., Ibanez, A.M., Rocheford, T.R., Goodman, M.M. and Buckler, E.S. (2004) Dissection of maize kernel composition and starch production by candidate gene association. *The Plant Cell* 16: 2719 -2733.

Yeh, F. C., Yang, R. C., Boyle, T. J. B., Ye, Z. H. and Mao, J. X. (1997) *Popgene, the User-Friendly Software for Population Genetic Analysis.* Edmonton, Molecular Biology and Biotechnology Centre, University of Alberta, Canada.

□□□

Forest Biotechnology in India © 2008
Editors: S.A. Ansari, C. Narayanan & A.K. Mandal, pp.109-120
Satish Serial Publishing House, Delhi

11

Biochemical and Molecular Markers for Genetic Diversity Poplar and Shisham

SALIL TEWARI, L. CHAUDHARY
S.K. UPADHYAY, PREKSHA AND RAJESH KAUSHAL

Forest Genetics Lab. (Agroforestry Section), Department of Genetics and Plant Breeding, G.B.Pant University of Agriculture & Technology, Pantnagar-263 145, Uttaranchal, India

ABSTRACT

Traditionally, the assessment of the genetic composition of germplasm have been conducted on the basis of morphological characters which is time consuming and blurred by environmental influences. Morphological markers lack the resolving power needed to identify specific genotypes. Estimation through biochemical markers may also be biased by the general consideration that only a minor portion of the genome is represented. Knowledge of genetic mechanisms underlying the variability in growth is almost invariably achieved through the use of molecular markers. In this study, nine clones of poplar could be easily distinguished on the basis of their protein profiles. Similarly, thirty clones of shisham have been distinguished on the basis of RAPD markers.

INTRODUCTION

Poplar (*Populus deltoides*) is one of the most important agroforestry tree species widely grown in states of north India. Poplars have found wide acceptability amongst the farming community for their fast growth and desirable phenological characters which make them convenient to raise in combination with agricultural crops. In general, poplars are highly heterozygous and outstanding good clones may be presumed to represent favorable heterotic conditions. Exploitation of genetic variability is essential for success of any breeding programme. Many genetically improved poplar clones have been identified in India. Before planting the poplar clones, it is necessary to confirm the clonal identity. Identification of clone on visual inspection\morphological description is complicated and time consuming. Alternatively, electrophoretic protein profiling procedures have been used in researches dealing with identification of genomes, cultivars, genera and genetic control of polymorphism and for tracing back the evolution of various group of plants. The advantage of protein profiling over other methods is that, it is highly repeatable and unique pattern for proteins can be obtained for each pure line, clone, cultivar or species under a given set of environmental conditions. Therefore, the present study was undertaken to identify the prominent clones through protein profile markers.

Shisham (*Dalbergia sissoo* Roxb.) is a multi purpose poplar timber tree species of Indian sub-continent well known for its timber, adaptation to adverse conditions, nitrogen fixing ability, good fuel wood and fodder. It is widely used in plantation programs under forestry, social and agroforestry systems. This species is well adapted to arid, barren and fragile sites. A large variability for growth characteristics exists in Shisham. Shisham genotypes also lack contrasting traits for comparative studies. DNA markers generated by random amplified polymorphic DNA (RAPD) have been proved to be useful for studying genetic and phylogenetic relationships in a wide range of organisms (Crowhurst *et al.*, 1991; Guthrie *et al.*, 1992). The main advantage of RAPD analysis is that no prior knowledge of DNA sequence is necessary in using the primers. RAPD markers have been used successfully for identification and classification of different genotypes of forest trees. Lin *et al.* (1994) differentiated clones of poplar and willow using four RAPD primers. Takata and Sirashi (1996) used RAPD markers for discriminating different cultivars of Japanese cedar.

MATERIALS AND METHODS

Protien profiling poplar clones

The present study was conducted on nine poplar clones viz., G_3, G_{148}, D_{121}, D_{163}, S_7C_1, S_7C_4, S_7C_8, S_7C_{15} and S_7C_{20} planted in the premises of College of Agriculture of G.B.Pant University, Pantnagar, U.S. Nagar, Uttaranchal.

Leaf samples were collected from different poplar clones in polythene bags. After removing the petiole and midrib, 0.5 g of leaf tissue from each clone was weighed and homogenized in 400 μl chilled extraction buffer using mortar and pestle, and incubated in refrigerator for 1 h. After incubation, the homogenized samples were centrifuged at 12000 rpm for 20 min. Later, the clear supernatant was carefully decanted into fresh Eppendorf tube and equal amount of sample buffer was added to it and mixed thoroughly. The samples were then kept at 37°C for 2 h and 35-45 μl of sample was loaded in each sample well in a polyacrylamide gel using a micro syringe. Initially, the electrophoresis run was carried out at 20 mA at normal room temperature. After the dye front crossed the stacking and separating gel interface the electrophoresis was continued at 30 mA. The run was continued till the tracking dye front migrated upto 1-1.5 cm from bottom of the separating gel. The entire run took around 6 h. After completion of electrophoresis the gel was stained overnight. After three times of destaining, gel was viewed and photographed under white light.

RAPD analysis of Shisham

Thirty genotypes of shisham collected from different parts of northern India and Nepal, and maintained in Germplasm Bank of Agroforestry Research Centre, Pantnagar, were selected for the study. Experiment was conducted at Forest Genetics Lab., Pantnagar and National Research Centre on DNA Fingerprinting, National Bureau of Plant Genetic Research, New Delhi. A list of genotypes with their accession numbers and origin are given in Table 1.

Table 1 : List of shisham germplasm accession and site of their selection

Shisham Accession No. (PS = Pantnagar Selection)	Site of selection
PS-49	Bajirgunj (Gonda, UP)
PS-39	Kanke (Ranchi, JH)
PS-15	Bajora-Kullu (HP)
PS-44	Mandar (Ranchi, JH)
PS-20	Bamimala (HP)

Contd...

Table 1. Contd...

PS-9	Guara (HP)
PS-22	Bamimala (HP)
PS-68	Brahmdev (Nepal)
PS-71	Paulkhera (Nainital, UA)
PS-50	Sisailbenjot (Gonda, UP)
PS-74	Gazipur (Gonda, UP)
PS-40	Kanke (Ranchi, JH)
PS-53	Bhainsadeh (Gonda, UP)
PS-7	Vikrampur (Ramnagar, UA)
PS-51	Jankinagar (Gonda,UP)
PS-28	Jwalamukhi (Kangra,HP)
PS-12	Solan (HP)
PS-36	Kukavali (Etawah, UP)
PS-45	Auriya (UP)
PS-41	Kanpur Dehat (UP)
PS-72	Tanakpur (UA)
PS-3	Haridwar (UA)
PS-5	Kaladungi (UA)
PS-48	Kanpur Dehat (UP)
PS-55	Balrampur (UP)
PS-75	Manikpur (Gonda, UP)
PS-8	Kaladungi (UA)
PS-70	Tanakpur (UA)
PS-69	Tanakpur (UA)
PS-26	Nauni (H.P.)

Collection and Preservation of Samples

Fresh leaves were collected, washed with distilled water, air-dried and cut into small pieces. Three grams of leaf tissue, wrapped in sterile aluminium foil, were immersed in liquid nitrogen for two minutes and transferred to deep freezer (-70^{0}C)

DNA extraction

Cetyltrimethylammonium bromide (CTAB) maxiprep DNA extraction protocol described by Saghai-Maroof *et al.* (1984), with certain modifications, was used for DNA isolation. Frozen leaf samples were ground into a fine powder using mortar and pestle and transferred to oakridge tube, containing extraction buffer and β-mercaptoethanol and of PVP. After gentle vortexing, samples were incubated at 60°C for 1 h. After slight vortexing, equal volumes of chloroformisoamyl alcohol (24:1) mixture and 1 ml of sodium acetate were added to the samples and swirled gently for five minutes to denature proteins. After centrifugation at 10,000 rpm for twenty minutes at 25°C, the aqueous phase was removed to a fresh tube. DNA was precipitated by adding 0.6 volumes of cold isopropanol, incubated at -20°C for 1 h and then centrifuged at 10,000 rpm at 10°C for 10 min. DNA pellet was washed with 70 per cent alcohol, dried at room temperature and dissolved in 1 ml of TE buffer (10 mM Tris-Cl and 1 mM EDTA). RNAse was added to extracted DNA and incubated for 1 h at 37°C to degrade RNA. The DNA was reprecipitated with with 0.1 volume of 3M sodium acetate and two volumes of ethanol (100%) at -20°C for 1 h, centrifuged and dried. Genomic DNA was dissolved in 50 ml TE buffer and quantified using fluorimeter.

RAPD analysis

PCR was carried out in 25 µl reaction mixture containing 25ng of genomic DNA, 200 mM each dNTPs, 1U Taq DNA polymerase enzyme, 1 x assay buffer, 1mM $MgCl_2$ and 1mM decamer primers (Operon Technologies Inc., USA) Primer codes and sequence codes are given Table 2. Amplification was performed in a Perkin - Elmer PCR system 9600 programmed for an initial denaturation of 4 min at 94°C, 40 cycles of denaturation at 94°C for 1 min, annealing at 35°C for 1 min and extension 72°C for 7 min. After amplification, 10 µl of PCR products were separated through electrophoresis in a 1.5% agarose gel in TBE buffer. DNA marker was used to estimate the molecular weight of the amplified products. Gel was visualized on UV transillluminator (302 nm) and photographed.

Table 2: Arbitrary decamer primers used for PCR amplification.

S. No.	Operon Code No.	Primer sequence (5′-3′)	GC content (%)
1	OPA 08	GTGACGTAGG	60
2	OPA 10	GTGATCGCAG	60
3	OPA 11	CAATCGCCGT	60
4	OPC 05	GATGACCGCC	70

Contd...

Table 2. Contd...

5	OPC 10	TGTCTGGGTG	60
6	OPF 09	CCAAGCTTCC	60
7	OPF 15	CCAGTACTCC	60
8	OPG 08	TCACGTCCAC	60
9	OPG 11	TGCCCGTCGT	70
10	OPH 01	GGTCGGAGAA	60
11	OPI 02	GGAGGAGA	50
12	OPI 19	AATGCGGGAG	60
13	OPJ 06	TCGTTCCGCA	60
14	OPK 10	GTGCAACGTG	60
15	OPL 05	ACGCAGGCAC	70
16	OPM 04	GGCGGTTGTC	70
17	OPN 05	ACTGAACGCC	60
18	OPP 06	GTGGGCTGAC	70
19	OPP 13	GGAGTGCCTC	70
20	OPQ 07	CCCCGATGGT	70
21	OPQ 20	TCGCCCAGTC	70
22	OPR 01	TGCGGGTCCT	70
23	OPR 05	GACCTAGTGG	60
24	OPS 03	CAGAGGTCCC	70
25	OPT 06	CAAGGGCAGA	60
26	OPQ 09	CCACATCGGT	60
27	OPQ 17	ACCTGGGGAG	70
28	OPV 05	TCCGAGAGGG	70
29	OPW 03	GTCCGGAGTG	70
30	OPZ 06	GTGCCGTTCA	60

Data analysis

DNA banding patterns generated by RAPD were manually scored as "1" for presence of an amplified band and "0" for absence. Polymorphic as well as monomorphic bands were used to generate Jaccard's similarity coefficient. The similarity coefficients were then used to manually construct a dendrogram by UPGMA (Unweighted Pair group Method with Arithmatical average).

RESULTS

Protein profiling of poplar clones

The comparison of electrophoretic banding patterns of the poplar clones under study clearly indicated that the protein band profiles were specific to each genotype. The formation of bands in the gel occurred in four zones, viz, I,II,III and IV. Zone I consisted of two bands, Ia and Ib. Band Ia was present in all the clones while band Ib was present only in clone S_7C_{20}. Zone II formed five bands: IIa, IIb, IIc, IId and IIe. Band IIa was present in clone D121, D163 and S_7C_4, band IIb was present in all clones, except S_7C_1 and S_7C_8; band IIc was observed in all clones under study, except S_7C_1, S_7C_8 and S_7C_{15}; band IId was observed in clones G3, G148 D121 and D163; whereas band IIe was observed only in clone D121. Zone III was characterized by four bands. Band IIIa was present in all clones except D163 while band IIIb was absent in all clones, except S_7C_4 and S_7C_{20}. Band IIIc was observed in clones G3 andG48. Band IIId was found in all clones except G48 and S_7C_1. Zone IV comprised two bands: IVa and IVb. Band IVa was absent in all clones, except G48 and D121 while band IV b was observed in all clones under study. Thus, on basis of variations in the number and intensity of bands, it is established that there are inherent differences among different genotypes for leaf protein profile, which can be useful in making identity of the different clones.

RAPD analysis of Shisham genotypes

RAPDs generated by the 30 primers with 30 shisham genotypes is presented in Table 3. 263 bands amplified with an average of 8.76 bands per primer of which 198 bands (75.28%) were polymorphic. RAPD profile of thirty genotypes with one primer, OPA 10 is presented in Fig. 1. Primers used in the study produced a number of unique markers which identified individual genotypes. Primer UPA-08 produced a 0.4 kbp size unique band for genotype PS-49. The same primer amplified a 2 kbp marker present only in genotypes PS-53, PS-70 and PS-26. A unique band of 3 kbp was amplified for PS-39 and PS-26 by primer OPA-10. PS-50, PS-51 and PS-36 exhibited a unique band of 0.45 kbp with the same primer. A unique fragment of 0.32 kbp in PS-20 was produced with this primer. The same primer amplified a fragment of 0.37 kbp size with PS-44 and PS-75. On amplification with primer OPA-11, genotype PS-20 gave distinct bands of 1.6, 1.1, 0.8, 0.7 and 0.6 kbp which gave unique banding pattern for the same, among thirty genotypes. A unique fragment of 0.21 kbp was amplified in case of PS-50 using OPC-05 primer. Primer OPC-10 generated a 2.2 kbp amplified product with genotype PS-20 and 0.9 kbp fragement with genotype PS-40 and PS-55. An amplified fragment of size 0.75 kbp was observed with PS-36, PS-3 and PS-75 using primer

OPG-08. Primer OPG-11 gave unique bands of size 2.7 and 3.0 kbp with genotype PS-71. It gave unique band of size 1.75 and 0.55 kbp with genotype PS-28 and PS-9, respectively. The same primer produced 0.35 kbp band with PS-68 and PS-13. A unique amplified product of 2.2 kbp was generated by PS-49 using OPH primer. It also generated a fragment of size 0.38 kbp shared by PS-7 and PS-69. Primer OPJ-06 produced amplified product of size 0.20 and 0.22 kbp, which was unique and shared only by PS-74 and PS-26. Genotype PS-28 produced a unique banding pattern with 3 bands of size 1.0, 0.8 and 0.6 kbp which differentiated it from rest of the 29 genotypes. Primer OPK-10 amplified 0.32 and 0.40 kbp product in PS-44 and PS-75, Primer OPL-05 produced a band of 3 kbp and produced the unique band of 0.61 kbp with PS-44 and PS-41. Two unique bands of size 1.19 and 0.82 kbp were also produced with PS-49 using OPM-04 primer. PS-53 and PS-8 generated a unique fragment of 0.89 kbp by use of the same primer. A fragment of 0.49 kbp was shared by PS-40, PS-12, PS-41 and PS-75 being generated by OPM-04. PS-39 and PS-40 generated 2.80 kbp unique fragment with primer OPN-05. This primer produced a fragment of 0.42 kbp being shared by PS-9, PS-45 and PS-55. Primer OPP-06 generated amplified product of size above 3.0 kbp which was unique to PS-68 and PS-5. It also produced unique band of 0.3 kbp with PS-20 and PS-41. A fragment of 0.5 kbp was generated with PS-74 using primer OPS-3. PS-68, PS-36 and PS-26 generated 1.01 kbp fragment with the same primer. Combination of two or more primers differentiated most of the genotypes. By using all 30 primers, all genotypes could easily be distinguished from each other. The primers OPG-11, OPM -04, OPN-05, OPL-05, OPA-08, OPA-10, OPK-10, OPP-06, OPS-03 and OPH-01 were found to give maximum number of unique and polymorphic markers.

Using data scored from 30 genotypes with thirty RAPD primers, Jaccard's similarity coefficients were calculated for all possible pairs of genotypes (Table 4). Similarity coefficients ranged between 0.502 (49.8%) to 0.776 (22.4%). Using Jaccard's similarity co-efficient values, UPGMA cluster analysis was performed. The dendrogram separated 30 genotypes into two major clusters. The first major cluster comprised 29 genotypes which were of Indian origin and the second cluster consisted of genotype from Nepal. The first main cluster consisted of two major sub-clusters. The first major sub-cluster included 23 genotypes, 19 from Uttar Pradesh (UP), three from Bihar and one from Himanchal Pradesh (HP). The second major sub-cluster comprised six genotypes, all from HP.

Table 3. RAPD markers of thirty genotypes of *shisham* detected by thirty arbitrary primers

S. No.	Primer Code	Total number of RAPD loci	Polymorphic loci	
			Number	Per cent
1	OPA 08	11	10	90.90
2	OPA 10	10	9	90.00
3	OPA 11	10	10	100
4	OPC 05	7	6	85.71
5	OPC 10	8	6	75.00
6	OPF 09	8	6	75.00
7	OPF 15	7	5	71.40
8	OPG 08	9	6	66.67
9	OPG 11	11	8	72.72
10	OPH 01	8	7	87.50
11	OPI 02	9	7	77.70
12	OPI 19	9	7	77.70
13	OPJ 06	10	8	80.00
14	OPK 10	10	8	80.00
15	OPL 05	13	12	92.30
16	OPM 04	9	6	66.67
17	OPN 05	15	13	86.66
18	OPP 06	11	10	90.90
19	OPP 13	6	4	66.66
20	OPQ 07	8	7	87.50
21	OPQ 20	10	7	70.00
22	OPR 01	10	4	40.00
23	OPR 05	7	3	42.80
24	OPS 03	7	5	71.42
25	OPT 06	7	3	42.80
26	OPQ 09	6	4	66.66
27	OPQ 17	5	4	80.00
28	OPV 05	8	5	62.50
29	OPW 03	6	3	50.00
30	OPZ 06	8	5	62.50

The first major subcluster with 23 genotypes further separated into two subclusters, SC-I with one genotype (PS-50) from Gonda, (UP) and SC-II with genotypes from Ranchi, Jharkhand, HP and UP.

Further study showed that SC-II included six different groups at different levels. First group comprised two genotypes, PS-71 from Nainital and PS-49 from Kanpur. Second group had four genotypes, PS-74, PS-51, PS-53 and PS-75, all from Gonda (UP). Third group was comprised of three genotypes, two from Etawah district and one from Balrampur of UP. Fourth group had four genotypes, PS-41 and PS-48 from Kanpur, PS-8 from Nainital and PS-26 HP. Fifth group comprised three genotypes, PS-72 and PS-5 from Nanital and PS-3 from Haridwar of Uttaranchal state. Sixth group of consisted of three genotypes, PS-7 from Ramnagar and PS-70 and PS-69 from Nainital district.

In general, groups I, II, III and IV comprised genotypes from UP while group III and IV consisted of genotypes from Uttaranchal. However, genotypes PS-71, PS-8 and PS-26 were exceptional.

The second major sub-cluster comprised six genotypes from HP which further bifurcated into two minor sub-clusters. First sub-cluster was formed of genotype PS-15 from Kullu while second sub-cluster was bifurcated in two groups, first group consisted of PS-20 and PS-22 from Baminala and the second group had PS-9 from Guara (HP).

DISCUSSION

Poplar and shisham lack distinctive morphological markers for identification of different genotypes. Being important tree species of plantation and natural forestry, poplar and shisham deserve critical studies at the molecular level to find out the molecular diversity among the different genotypes. Different poplar clones could be characterized through protein profiling. These protein profiles can be used for identification of the above poplar genotypes. Similarly, thirty genotypes of shisham were differentiated using thirty arbitrary decamer primers.

UPGMA cluster analysis could readily separate thirty genotypes of shisham into different groups. PS-68 from Nepal separated from Indian genotypes. Genotypes from hill sate of HP formed separate clusters, with exception of PS-26. Accessions from Bihar state aggregated in a group showing close proximity to each other. Four genotypes (PS-74, PS-51, PS-53 and PS-75) from Gonda, UP, two (PS-36, PS-45) from Etawah, UP, three genotypes (PS-72, PS-5 and PS-3) from Nainital, Uttaranchal, formed separate group.

The above study indicated that genotypes originating in close proximity to each other are more closely related at the genetic level than the rest. Gato *et al.* (1997) drew the same inference while classifying *Rhus succedanea* L. genotypes using RAPD markers. Grouping may be explained on following basis: (a) Certain genotypes had migrated to other ecological niche long time ago and acclimatized excellently, (b) During natural or man made selections, genotypes with similar genetic constitution were selected at different places which are now falling in the same group and (c) Both eco-geographical zones are having broad genetic variability. Earlier studies have also shown that morphological markers do not necessarily reflect the genetic markers as recorded by Sale *et al.* (1996) in eucalyptus and Krahl *et al.* (1993) in red maple. The present findings showed that RAPD markers were able to discriminate genotypes of poplar and shisham into distinct groups in the absence of distinct morphological traits. RAPD markers provide a better coverage of the whole genome, resulting in a better estimate of genetic relationship (Smith and Smith, 1989).

References

Crowhurst, R.N., Nowthorne, B.T., Rikkerink, E.H.A. and Templeton, M.D. (1991) Differentiation of *Fusarium solani* f.sp. *cucurbitae* race 1 and 2 by random amplification of polymorphic DNA. *Curr. Genet.* 20: 391-396.

Gato, S., Watanabe, A. and Ikeda, K. (1997) Use of RAPD markers for cultivar identification in *Rhus succedanea* L. *J. Japanese Forestry Soc.* 79: 229-233.

Guthrie, P.A . I., Magill, C.W., Frederiksen, R.A. and Odvody, G.N. (1992) Random amplified polymorphic DNA markers: A system for identifying and differentiating isolates of *Colletrotrichum graminicola. Phytopathol.* 82: 832-835.

Krahl, K.H., Dirr, M.A., Halward, T.M., Kochert, G.D. and Randle, W.M. (1993) Use of single primer DNA amplifications for the identification of red maple (*Acer rubrum* L.). *J. Env. Hort.* 11: 89-92.

Lin, D., Hubbes, M., Zsuffa, L. and Dietrichson, J. (1994) Differentiation of poplar and willow clones using RAPD fingerprints. *Tree Physiol.* 14: 1097 – 1105.

Sale, M.M., Potts, B.M., West, A.K. and Reid, J.B. (1996) Molecular differentiation within and between *Eucalyptus risdonii*, *E. amygdalina* and their hybrids using RAPD markers. *Aust. J. Bot.* 44: 559-569.

Saghai-Maroof, M.A., Soliman, K.m., Jorgensen, R.A. and Allard, R.W. (1984) Ribosome DNA spacer length polymorphism in barley : Mendelian inheritance, chromosomal location and population dynamics. *Proc. Nat. Acad. Sci., USA* 81: 8014-8018.

Smith, J.S.C. and Smith, O.S. (1989) The description and assessment of distance between inbred lines of maize. II. The utility of morphological, biochemical and genetic descriptors and a scheme for testing of distinctveness between inbred lines. *Maydica* 34: 151-161.

Takata, K. and Sirashi, S. (1996) Discrimination of Japanese cedar (*Cryptomeria japonica* D.Don) cutting cultivars in Kyushu region with RAPD markers. Bull Kyushu Univ. *Forests* 75: 1-14.

□□□

Forest Biotechnology in India © 2008
Editors: S.A. Ansari, C. Narayanan & A.K. Mandal, pp.121-136
Satish Serial Publishing House, Delhi

12

Allozyme Variation in Populations of *Pinus Roxburghii* Sargent in Western Himalayas

RAJESH SHARMA, D.K. KHURANA AND GUNAR SCHÜTE

Himalayan Forest Research Institute, Conifer Campus
Panthaghati, Shimla-171009, India

ABSTRACT

Genetic variability in populations of *Pinus roxburghii* the most valued pine of Western Himalayas, was investigated. Eight enzyme systems (IDH, MDH, GDH, SKDH, MNR, LAP, 6PGDH, and PGM) were assayed using starch gel electrophoresis. Intra-specific variation was studied to determine the genetic variation between seven populations. Of the 16 presumable gene loci 8 were polymorphic with 2-4 alleles per locus. The pattern of distribution of genetic diversity and differentiation among the populations, despite altitudinal differences, suggest common descent of the populations.

Keywords : *Pinus roxburghii*, isoenzymes, populations, diversity, genetic distance

INTRODUCTION

Genetic variation is the basis for adaptation and survival of living organisms under changing environmental conditions. Forest tree species owing to generally predominantly out crossing possess a relatively high level of genetic variation, and a high degree of heterozygosity. Although genetic surveys of tree species have tended to estimate low levels of variation (Ledig, 1986; Bergman, 1991), the genetic diversity in trees is generally twice than that in herbaceous plants.

The genetic diversity maintained by plant species and its distribution within and among populations are often associated with certain suites of ecological and life history traits (Hamrick *et al.*, 1979, 1992; Loveless and Hamrick, 1984; Hamrick and Godt, 1989). For example, long-lived woody plants tend to maintain more genetic variation at both species and population levels, but have less genetic differentiation among populations than non-woody species. These trends are exemplified by the gymnospermous species, especially coniferous trees.

Among the various valuable natural resources of conifers, pines constitute one of the most divergent and economically important groups of species in the Himalayas. *Pinus roxburghii* Sargent, commonly known as chir pine, is the most important among the six indigenous pines of India, valued for its timber and oleoresin. This pine is also the most widely distributed natural pine over a long strip of 3200 km from east to the west in the monsoon belt of the outer Himalayas from Bhutan to Pakistan. It occurs at an elevation of 450-2300 m, with optimum growing range of 900 to 1800 m (Dogra, 1985).

1. University of Horticulture & Forestry College of Forestry, Nauni, Solan, India

2. Institute of Forest Genetics & Tree Breeding, Escherode, Germany

The genetic improvement of chir pine has remained comparatively neglected, and mostly restricted to provenance testing because of its easy regeneration in managed stands. Provenance trials conducted under Indo-Danish project during 1979 revealed significant variation in growth rates between localities (Uniyal and Rawat, 1988). Other studies showed intra-specific variability in morphotypes (Khosla *et al.*, 1988) and in selected plus trees/seed stands of the species (Sagwal, 1978). In comparative studies on progenies of two types of trees, i.e. high resin yielders and trees with good growth characters, high resin yielders were found to be inferior in growth characters (Sehgal *et al.*, 1987).

The provenance tests based on quantitative measurements often does not give real picture about the genetic variation as the measurements are subjected to environmental influences. Hence, the traditional procedures need to be supplemented by the modern techniques such as isozyme analysis for detecting systematic differences between populations of a species.

Unlike other pine species, however, not much work has been done on isozyme analysis in chir pine and studies are only confined to two investigations using 12 and 8 chir pine populations from Pakistan and India, respectively (Hussain, 1995; Sharma *et al.* 2002). Therefore, keeping in view the results of the past studies especially provenance tests (Uniyal and Rawat, 1988) and progeny tests (Sehgal *et al.*, 1987), the present study incorporates 7 distinct populations identified based upon growing stock, soil, forest types and eco-climatic conditions to measure distribution of genetic variation through allozymes. The work was done at the Institute of Forest Genetics and Tree Breeding Escherode, Germany.

MATERIAL AND METHODS

Seven geographically distinct populations formed the study material for the present investigation (Table 1). Open pollinated seeds

Table 1: Details of *Pinus roxburghii* populations used for isozme analysis.

S.No	Populations	Altitude (m)	Latitude	Longitude	Forest Type*	Soil Type	Growing stock (1000 m^3)
1	Hamirpur	800	31^0 36' N	76^0 30' E	9/C1a Shiwalik chir pine	Gravelly loam	2135
2	Solan	1500	30^0 56' N	76^0 57' E	9/C1a Shiwalik chir pine	Fairly Deep, loam	1536
3	Banethi	1200	30^0 51' N	76^0 48' E	9/C1a Shiwalik chir pine	Clayey loam	1205
4	Nurpur	750	30^0 55' N	76^0 38' E	9/C1a Shiwalik chir pine	Sandy loam	2648
5	Shimla	2100	31^0 07' N	77^0 08' E	9/C1b Himalayan chir pine	Shale and lime stone	507
6	Karsog	1200	31^0 09' N	76^0 42' E	9/C1b Himalayan chir pine	Fertile sandy loam	697
7	Jogindernagar	1300	31^0 46' N	77^0 41' E	9/C1a Shiwalik chir pine	Loam	1247

*Champion & Seth (1968)

were harvested from 20 trees from each population and kept separately maintaining their identity for isozyme analysis. Ten each of megagametophyte and embryo samples were assayed for 8 enzyme systems.

Horizontal starch gel electrophoresis

Methods of Shaw and Prasad (1970), Conkle *et al.* (1982) and Cheliak and Pitel (1984) were slightly modified for isozyme analysis. Macrogametophyte tissues and embryos were isolated separately from the seeds and homogenized in 0.1 M Tris HCL buffer (pH 7.5). Homogenates were subjected to horizontal starch gel electrophoresis with composition of starch (15%), saccharose and gel buffer. All the isozyme systems were analysed using Tris-Citrate System.

Statistical analysis

The data were analysed using GSED software (Gillet, 1994). The measures of genetic diversity included percentage of polymorphic loci, average number of alleles per locus, average number of alleles per polymorphic locus, allele and genotype frequencies, heterozygosity, allelic diversity and differentiation and genetic distance. Tests of homogeneity (G-test and chi square) were used for calculating allele and genotype frequencies.

RESULTS AND DISCUSSION

Isozyme analysis

For 8 enzyme systems studies, 16 presumable loci were identified, out of which 8 were polymorphic. However, one polymorphic locus namely MDH-B was not included in the analysis due to its non-stability. The details of alleles for different putative enzyme gene loci are given in Table 2.

The genetic control of isozymes was postulated from the banding patterns observed in megagametophytes and embryos of trees of populations. In total, 16 zones of activity were recorded for the 8 enzyme systems assayed as described in following sections.

Isocitrate dehydrogenase (IDH)

Two zones of activity (IDH-A and IDH-B) were observed for IDH. The faster migrating zone (IDH-A) stained heavily in comparison to the slower one (IDH-B) and was monomorphic. At locus IDH-B, two

Table 2: Enzymes, gene loci and their allelic variants in *P. roxburghii*.

S. No.	Enzyme	Gene loci	No. of alleles per locus
1	Glutamate dehydrogenase E.C. 1.4.1.2	GDH-A	1
2	Isocitrate dehydrogenase E.C. 1.1.1.42	IDH-A IDH-B	1 2
3	Leucine aminopeptidase E.C. 3.4.11.1	LAP-A LAP-B	2 1
4	Malate dehydrogenase E.C. 1.1.1.37	MDH-A MDH-B MDH-C MDH-D	1 - 2 1
5	Menadione reductase E.C. 1.6.99.2	MNR-A MNR-B MNR-C	1 2 1
6	Phosphoglucomutase E.C. 2.7.5.1	PGM-A	3
7	6-Phosphogluconate dehydrogenase E.C. 1.1.1.41	6PGDH-A 6PGDH-B	4 1
8	Shikimic acid dehydrogenase E.C. 1.1.1.25	SKDH-A	3

phenotypes encoded by two faint bands were found for the endosperm tissue. In embryos, in addition to the phenotypes observed for endosperms, a third phenotype with four bands was also observed. This observation is in agreement with the earlier results of Hussain (1995) and Sharma *et al.* (2002) in the species.

Shikimic acid dehydrogenase (SKDH)

For SKDH, one zone of activity with three single banded phenotypes was identified both for endosperm and embryos. In addition, three double-banded phenotypes were also noticed in embryos. Appearance of two banded phenotypes in heterozygous embryos and only single banded phenotype in endosperms indicated the monomeric structure of SKDH. Out of the three alleles recorded for SKDH, allele 2 occurred frequently followed by allele 3 and allele 1 Hussain (1995) and Sharma *et al.* (2002) reported a second weak

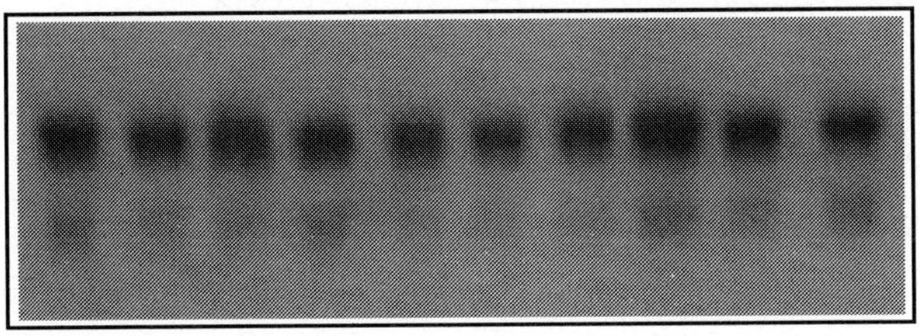

A1 B2   A11 B11   A1 B1   A11 B22   A1 B1   A11 B11   A1 B1   A11 B12   A11 B11   A11 B11

staining monomeric locus in *P. roxburghii*; however, this locus was not observed in the present study. Hertel and Maurer (1998) reported a very weak staining or sometimes absence of staining of locus B for SKDH in *Pinus sylvestris*.

6-Phosphogluconate dehydrogenase (6-PGDH)

Two zones of activity with almost similar staining intensities were observed for 6PGDH. 6PGDH-A had four allozymes whereas 6PGDH-B was monomorphic. The allozyme 6PGDH-A2 was present in all the samples followed by A1. Allozyme 6PGDH-A3 with 2.9 % mean allele frequency and A4 with 1.98 % mean allele frequency were observed in all the populations. Hussain (1995) and Sharma *et al.* (2002) also reported two gene loci for this system but found three alleles for the variant locus 6PGDH-A in contrast to four found in the present study.

A22 A2 A22 A12 A22 A2 A12 A2 A22 A2

Malate dehydrogenase (MDH)

Sharma and Wuehlisch (1998) described two detailed staining methods for interpretation of MDH and have categorically mentioned that interpretation of genetic control of MDH is difficult due to a gene duplication of a certain gene locus and formation of inter locus heterodimeric hybrid band. This enzyme system is controlled by four gene loci. In the present study, MDH-D was found to be monomorphic. Due to overlapping of alleles, locus MDH-A was found to be staining intensively but did not show any variation and hence, interpreted as monomorphic. MDH-B was polymorphic but it was not stable and hence, not included in the analysis. An interlocus hybrid band between MDH-B and MDH-C was also observed as a specific feature in MDH,

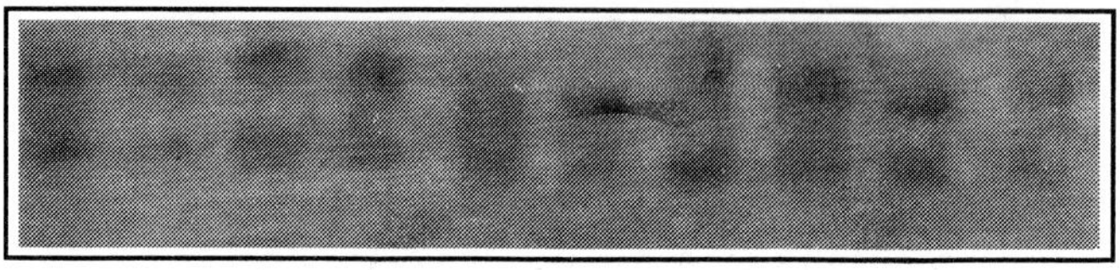

A2	A22	A1	A12	A2	A22	A1	A12	A22	A12
B1	B11	B1	B11	B1	B11	B1	B11	B11	B11

which was commensurate with the observations of El-Kassaby (1981), Thormann and Stephan (1993) and Sharma *et al.* (2002). MDH-C was variable with two allozymes, out of which C2 was more frequent accounting for 60 per cent of the gene locus.

Leucine aminopeptidase (LAP)

This system exhibited two gene loci with monomorphic structure. Minor polymorphism was observed for LAP-A. LAP-A showed the presence of one allozyme with a single band. The absence of a band in the LAP-A zone and detection of a band at LAP-B indicated null allele at LAP-A locus. In this zone, the allozyme A1 was observed frequently but the allozyme A0 was found in all the populations except population 1. Presence of null allele at LAP-A locus in conifers is a common observation and has also been reported by Hussain (1995) and Sharma *et al.* (2002) in *P. roxburghii*, by Konnert (1995) in *Picea abies* and by Hertel (1997) in *P. sylvestris*.

A1	A11	A1	A11	A1	A11	A1	A11	A11	A11
B2	B22	B2	B22	B2	B22	B2	B 22	B12	B22
C2	C12	C2	C22	C2	C12	C2	C12	C11	C12
D1	D11	D1	D11	D1	D11	D1	D11	D11	D11

Phosphoglucomutase (PGM)

The enzyme system was represented by one gene locus with monomeric structure. At locus PGM-A, which was found to stain relatively quickly, three alleles were found with A2 being the most abundant variant followed by alleles A1 and A3. Hussain (1995) and Sharma *et al.* (2002) described two gene loci with two variants for PGM-A in *P. roxburghii*. However, in *Pinus sylvestris* Goncharenko *et al.* (1994) observed two PGM loci with three variants at PGM-A. Hussain (1995) and Sharma *et al.* (2002) also reported the presence of null allele for this locus, which was not observed in the present study. These differences for PGM-A might be either due to the detection system used or due to differences in the genetic make up of the experimental populations. Hertel and Maurer (1998) also reported single gene loci for PGM in *Pinus sylvestris* and stressed that the second gene loci for the system, if present, could be due to some unspecific reactions.

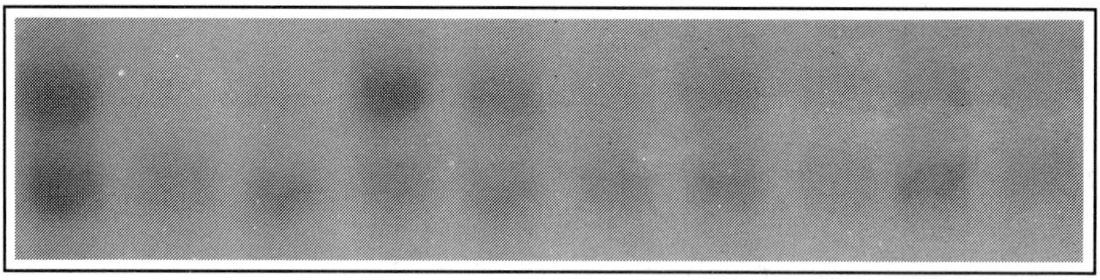

A1 B1   A00 B11   A0 B1   A11 B 11   A1 B 1   A00 B11   A1 B1   A00 B11   A11 B11   A11 B11

Glutamate dehydrogenase (GDH)

Only one zone of activity was observed for GDH. In all megagametophyte and embryo extracts, no variation was found as only a single allelic variant encoded by a sharp band was noticed. Hussain (1995) and Sharma *et al.* (2002) also reported single gene locus for *P. roxburghii*. A single zone of activity for GDH has also been reported in other conifers (Guries and Ledig, 1978; Adams and Joly, 1980; Yeh and El-Kassaby, 1980; Hertel and Maurer, 1998).

A1 A22 A2 A22 A2 A23 A2 A23 A22 A23

Menadione reductase (MNR)

Gels stained for MNR showed three zones of activity, viz., MNR-A, MNR-B and MNR-C. MNR-A was stained intensively compared to MNR-B and MNR-C. MNR-A and MNR-C were monomorphic whereas MNR-B was found with two phenotypes. Two allozymes of MNR-B were uniformly distributed, except for one population where the allele 1 was more frequent than the allele 2. Hussain (1995) and Sharma *et al.* (2002) observed polymorphism for MNR-A, which was not observed in the present study.

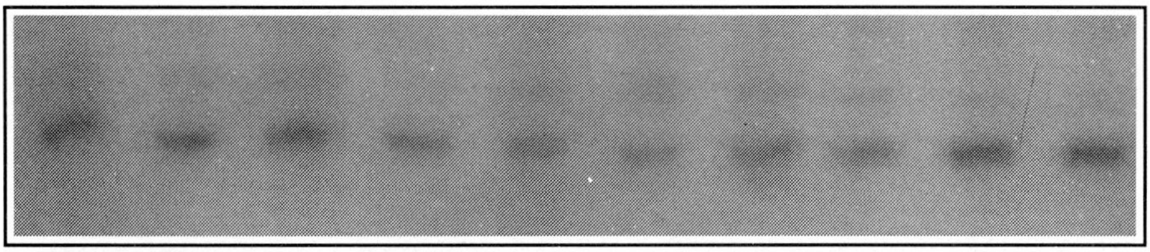

A1 A11 A1 A11 A1 A11 A1 A11 A11 A11

Allele frequency distribution

The frequencies of the respective alleles of the polymorphic loci in different populations are given in Table 3. The frequency distribution of the alleles showed differences among different populations. Out of 16 alleles for 6 polymorphic gene loci, PGM-A3 was rare with minimum (<1%) frequency in all the populations. Similarly, the allele 6PGDH-A3 with minimum (<1%) frequency in two populations. The allele 6PGDH-A4 had less than 1 % allele frequency for two populations and varied from 1.8 to 3.5 % in rest of the populations.

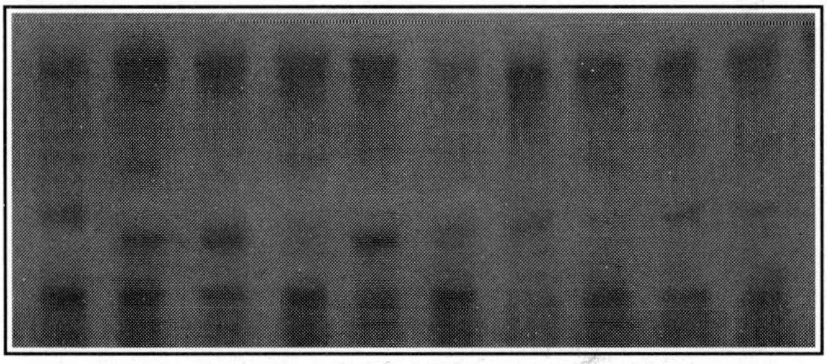

A1 A11 A1 A11 A1 A11 A1 A11 A11 A11
B1 B22 B2 B12 B2 B12 B1 B11 B11 B11
C1 C11 C1 C11 C1 C11 C1 C11 C11 C11

Allele frequency for SKDH ranged from 1.2 (population 2) to 17.3 % (population 4). For 6PGDH-A1, allele frequency ranged from 4.7 to 20.8 %. The allele frequency for PGM-A1 ranged from 1.9 to 20.2 % but absent in population 6. MNR-B, MDH-C and IDH-B exhibited high levels of polymorphism among different populations. MNR-B11 allele frequency of ranged from 48.8 to 68 %. Further analysis of the data revealed that PGM A3 is a rare allele as it was observed with very low frequency (<1%) in all the populations.

Genotype frequency distribution

A total of 29 genotypes for 6 polymorphic loci were observed in the populations (Table 4). The genotypes 6PGDH A34 and 44 were

Table 3: Allele frequencies for 7 populations of *Pinus roxburghii*.

Gene loci	Allele	Populations							Mean
		1	2	3	4	5	6	7	
IDH-B	1	69.4	63.7	61.2	79.7	71.9	71.6	78.6	70.9
	2	30.6	36.3	38.8	20.3	28.1	28.4	21.4	29.1
SKDH-A	1	11.6	1.2	17.1	17.3	4.2	3.8	7.0	8.9
	2	77.7	86.0	64.7	75.5	78.6	69.5	72.8	74.9
	3	10.7	12.8	18.2	7.2	17.2	26.8	20.2	16.2
6PGDH-A	1	17.3	13.1	8.6	8.2	6.3	20.8	4.7	11.3
	2	79.4	79.8	85.8	87.4	87.4	75.7	91.4	83.8
	3	1.5	5.2	4.7	3.5	4.0	0.9	0.5	2.9
	4	1.8	1.9	0.9	0.9	2.3	2.6	3.5	2.0
MDH-C	1	37.6	35.6	40.5	48.6	46.3	33.7	32.9	39.3
	2	62.4	64.4	59.5	51.4	53.7	66.3	67.1	60.7
PGM-A	1	20.2	2.8	15.8	2.1	1.9	-	1.9	6.4
	2	79.5	97.0	83.7	97.4	97.4	99.1	97.6	93.1
	3	0.3	0.2	0.5	0.5	0.7	0.9	0.5	0.5
MNR-B	1	48.8	51.6	51.2	50.2	57.0	55.6	68.0	54.6
	2	51.2	48.4	48.8	49.8	43.0	44.4	32.0	45.4

Table 4: Genotype frequencies for 7 populations of *Pinus roxburghii*.

Gene loci	Genotype	Populations							Mean
		1	2	3	4	5	6	7	
IDH-B	1x1	51.9	42.8	38.8	65.9	54.0	55.0	66.2	53.5
	1x2	35.11	41.9	44.9	27.6	35.8	33.2	24.9	34.8
	2x2	3.0	15.3	16.4	6.5	10.2	11.8	8.9	11.7
SKDH-A	1x1	1.2	-	9.8	9.1	1.4	0.9	2.8	3.6
	1x2	20.7	2.3	13.5	16.3	5.1	5.6	7.9	10.2
	1x3	-	-	0.9	-	0.5	-	0.5	0.3
	2x2	57.4	74.9	46.3	62.1	65.1	48.8	54.0	58.3
	2x3	20.1	20.0	23.4	10.6	21.9	35.8	29.7	23.1
	3x3	0.6	2.8	6.1	1.9	6.0	8.9	5.1	4.5
6PGDH-A	1x1	10.3	7.0	2.3	1.4	2.3	8.0	-	4.5
	1x2	13.9	9.9	11.6	12.6	7.4	22.6	9.3	12.5
	1x3	-	1.9	0.5	0.9	0.5	0.5	-	0.6
	1x4	-	0.5	0.5	-	-	2.4	-	0.5
	2x2	69.7	69.0	74.9	78.0	78.6	62.3	83.3	73.6
	2x3	1.8	8.5	8.8	4.3	6.0	1.4	0.9	4.5
	2x4	3.6	3.2	1.4	1.9	4.2	2.8	6.0	3.3
	3x3	0.7	-	-	0.9	0.5	-	-	0.3
	3x4	-	-	-	-	0.5	-	-	0.1
	4x4	-	-	-	-	-	-	0.5	0.1
MDH-C	1x1	17.0	19.1	28.0	33.8	31.2	19.1	18.3	23.8
	1x2	41.2	33.0	25.1	29.6	30.2	29.3	29.1	31.1
	2x2	41.8	47.9	46.9	36.6	38.6	51.6	52.6	45.1
PGM-A	1x1	17.4	-	13.0	1.4	-	-	-	4.6
	1x2	5.6	5.5	5.6	1.4	3.7	-	3.7	3.6
	2x2	76.4	94.0	80.5	96.3	94.9	98.1	95.4	90.8
	2x3	0.6	0.5	0.9	0.9	1.4	1.9	0.9	1.0
MNR-B	1x1	25.5	29.9	27.9	27.0	32.1	34.4	43.9	31.5
	1x2	46.6	43.5	46.5	46.5	49.8	42.3	48.2	46.2
	2x2	27.9	26.6	25.6	26.5	18.1	23.3	7.9	22.3

rare and found only in populations 5 and 7. The genotype 6PGDH - A33 was found in three populations (1, 4 and 5) with less than 1 % genotypic frequency. The genotype SKDH-A13 was observed with less than 1 % frequency in populations 3, 5 and 7. The population 7 showed comparatively high values for IDH-B11. The differences in genotypic frequencies for SKDH-A11 varied from 0.9 to 9.8 %. Maximum numbers (26) of genotypes were observed in the populations 3 and 5 and minimum (23) in populations 2, 6 and 7.

The heterozygosity among the populations ranged from 0.25 to 0.45 for IDH-B, 0.22 to 0.41 for SKDH-A, 0.16 to 0.3 for 6PGDH-A, 0.25 to 0.41 for MDH-C and 0.02 to 0.07 for PGM-A. The mean heterozygosity ranged from 0.25 to 0.32 for populations 4 and 1 respectively (Table 5).

Table 5: Details of heterozygosity at 6 polymorphic loci in populations of *Pinus roxburghii.*

Gene loci	Populations							Range
	1	2	3	4	5	6	7	
IDH-B	0.352	0.419	0.449	0.276	0.358	0.332	0.249	**0.249-0.449**
SKDH	0.409	0.223	0.379	0.269	0.274	0.413	0.381	**0.223-0.413**
6PGDH-A	0.194	0.239	0.228	0.196	0.186	0.297	0.163	**0.163-0.297**
MDH-C	0.412	0.330	0.251	0.296	0.302	0.293	0.291	**0.251-0.412**
PGM	0.062	0.060	0.065	0.023	0.051	0.019	0.047	**0.019-0.065**
MNR-B	0.467	0.435	0.465	0.465	0.498	0.423	0.481	**0.423-0.498**
Mean (H_a)	**0.316**	**0.284**	**0.306**	**0.254**	**0.278**	**0.296**	**0.269**	**0.254-0.316**

Allelic differentiation

The gene pool differentiation among the populations ranged from 30.8 to 41.8 %. Population 3 was the most differentiated followed by population 1 (Table 6). Population 7 was the least differentiated with a value of 30.8 %. Large differences were observed for allelic differentiation between and within the gene loci among the populations. The mean differentiation (d) values for the 6 gene loci varied from 0.11 (PGM) to 0.48 (MNR-B).Wide ranges of differentiation values were recorded for different populations within gene loci e.g., PGM (0.01 for population 6 to 0.32 for population 1), 6PGDH-A (0.16 for population 7 to 0.38 for population 6) and SKDH (0.24 for population 2 to 0.52 for population 3). Overall, a high amount of allelic differentiation among populations was reflected by the mean gene pool value (d = 35.9) which indicated that populations differed by 35.9 % of the effective number of alleles.

Table 6: Allelic differentiation (values in parenthesis) in populations of *Pinus roxburghii.*

Gene loci	Populations							Mean (δ)
	1	2	3	4	5	6	7	
IDH-B	1.737 (0.426)	1.860 (0.463)	1.904 (0.476)	1.479 (0.325)	1.679 (0.405)	1.686 (0.408)	1.506 (0.337)	1.693 (0.406)
SKDH-A	1.589 (0.372)	1.321 (0.244)	2.078 (0.520)	1.653 (0.396)	1.540 (0.352)	1.799 (0.445)	1.737 (0.425)	1.673 (0.393)
6PGDH-A	1.513 (0.340)	1.521 (0.344)	1.340 (0.255)	1.296 (0.229)	1.298 (0.230)	1.621 (0.384)	1.192 (0.162)	1.397 (0.278)
MDH-C	1.884 (0.471)	1.846 (0.459)	1.931 (0.483)	1.998 (0.501)	1.989 (0.498)	1.808 (0.448)	1.790 (0.442)	1.892 (0.472)
PGM-A	1.486 (0.328)	1.062 (0.059)	1.378 (0.275)	1.053 (0.050)	1.053 (0.050)	1.019 (0.019)	1.048 (0.046)	1.157 (0.118)
MNR-B	1.999 (0.501)	1.998 (0.501)	1.999 (0.501)	2.0 (0.501)	1.962 (0.491)	1.975 (0.495)	1.771 (0.436)	1.957 (0.489)
Gene Pool	**1.681 (0.406)**	**1.525 (0.345)**	**1.716 (0.418)**	**1.499 (0.334)**	**1.508 (0.338)**	**1.576 (0.367)**	**1.444 (0.308)**	**1.628 (0.359)**

Genotypic differentiation

The pattern of genotypic differentiations in the populations (Table 7) was similar to that of allelic differentiation and varied from 0.44 (population 7) to 0.56 (population 3). The mean genotypic differentiation (d) among the 6 gene loci ranged from 0-16 (PGM-A) to 0.63 (MNR-B) with mean gene pool value of 0.59.

Table 7: Genotypic differentiation (values in parenthesis) in populations of *Pinus roxburghii.*

Gene loci	Populations							Mean (δ)
	1	2	3	4	5	6	7	
IDH-B	2.442 (0.594)	2.619 (0.621)	2.643 (0.625)	1.944 (0.488)	2.326 (0.573)	2.346 (0.576)	1.968 (0.494)	2.326 **(0.567)**
SKDH-A	2.426 (0.591)	1.661 (0.400)	3.329 (0.703)	2.319 (0.571)	2.091 (0.524)	2.653 (0.626)	2.568 (0.613)	2.435 **(0.575)**
6PGDH-A	1.932 (0.485)	2.002 (0.503)	1.716 (0.419)	1.594 (0.374)	1.589 (0.372)	2.237 (0.556)	1.417 (0.296)	1.783 **(0.429)**
MDH-C	2.677 (0.630)	2.667 (0.628)	2.767 (0.642)	2.978 (0.667)	2.963 (0.666)	2.572 (0.614)	2.533 (0.608)	2.736 **(0.636)**
PGM-A	1.621 (0.385)	1.129 (0.115)	1.498 (0.334)	1.078 (0.073)	1.109 (0.099)	1.038 (0.037)	1.098 (0.090)	1.224 **(0.162)**
MNR-B	2.776 (0.644)	2.863 (0.654)	2.780 (0.643)	2.782 (0.644)	2.607 (0.619)	2.843 (0.651)	2.321 (0.572)	2.710 **(0.632)**
Gene Pool	**(0.555)**	**(0.487)**	**(0.562)**	**(0.470)**	**(0.476)**	**(0.510)**	**(0.446)**	**(0.595)**

Test of homogeneity for distribution of allele frequencies

Significant differences were observed among the allelic frequencies of the seven populations (Table 8). G-test and c^2 tests were significant for all the polymorphic loci. The values for G-test and c^2 tests for genotypic frequency distributions among seven populations were also significant for all the gene loci tested.

Table 8: The homogeneity tests for allelic and genotypic frequencies.

S. No.	Locus	Allelic frequencies		Genotypic frequencies	
		G-Test	Chi square	G-Test	Chi square
1	IDH-B	59.599***	59.162***	59.076***	58.074***
2	SKDH-A	213.456***	204.530***	199.026***	194.054***
3	6PGDH-A	133.231***	130.299***	160.194***	160.946***
4	MDH-C	40.204***	40.348***	44.218***	44.967***
5	PGM-A	241.074***	269.321***	176.958***	180.403***
6	MNR-B	44.316***	43.460***	50.967***	45.781***

***significant at 0.1%

Genetic Distance

The matrix of the genetic distance for seven populations is given in Table 9. Genetic distance values ranged from 0.05 between populations 4 and 5 to 0.12 between populations 1 and 7. Out of 21 pair wise combinations, 24 % showed values of the genetic distance more than 0.10 whereas and 38 % showed values less than 0.08 and the remaining 38 % showed intermediate values. The population 7 maintained maximum genetic distance with rest of the populations.

Table 9 : Genetic distance for 7 populations of *Pinus roxburghii*.

Population No.	1	2	3	4	5	6
1	-					
2	0.071	-	-			
3	0.068	0.081	-			
4	0.094	0.091	0.090	-		
5	0.093	0.068	0.088	0.054	-	
6	0.089	0.069	0.107	0.107	0.067	-
7	0.124	0.102	0.123	0.091	0.071	0.076

Based on population parameters, tendency of grouping was observed between the populations. Taking allelic diversity and differentiation into consideration, population 7 remained separated from rest of the populations. Population 1 and 3 made one group and populations 2, 4, 5 and 6 another group. The same tendency was observed for genotypic differentiation. The genetic distance values further supported this grouping as population 7 showed relatively higher genetic distance with the remaining populations whereas the remaining populations showed less genetic distance among themselves. The tendency of grouping among different populations irrespective of their altitudinal differences suggests a common descent of the populations.

References

Adams, W.T. and Joly, R.J. (1980) Genetics of allozyme variants in Loblolly pine. *J. Hered.* 71: 33-40.

Bergmann, F. (1991) Isozyme Gene Markers. In: *Genetic Variation in European Populations of Forest Trees* (eds Mueller-Starck, G. and Ziehe, M.) J.D. Sauerlander' Verlag, Frankfurt am Main, pp. 67-78.

Cheliak, W.M. and Pitel, J.A. (1984) Genetic control of allozymes variants in mature tissues of White spruce trees. *J. Hered.* 75: 34-40 .

Conkle, M.T., Hodgskiss, P.D., Hunnally, L.B. and Hunter, S.C. (1982) Starch gel electrophoresis of conifer seed: A laboratory manual.USDA Forest Service General Technical Report PWS-64, 18 p.

Dogra, P.D. (1985) Conifers of India and their wild gene resources in relation to tree breeding. *Indian For.* 111: 935-955.

EL-Kassaby, Y.A. (1981) Genetic interpretation of malate dehydrogenase isozymes in some conifer species. *J. Hered.* 72: 451-452.

Gillet, E. (1994) Genetic structures from electrophoretic data (GSED). Version 1.0. Abteilung fur Forstgenetik und Forstpflanzenzuchtung, Universitat Gottingen, Busgenweg 2, D- 37077 Gottingen, Germany.

Goncharenko, G.G., Silin, A.E., and Padutov, V.E. (1994) Allozyme variation in natural populations of Eurasion pines. III. Population structure, diversity, differentiation and gene flow in central and isolated populations of *Pinus sylvestris* L. In eastern Europe and Siberia. *Silva. Genet.* 43: 119-132.

Guries, R.P. and Ledig, F.T. (1978) Inheritance of some polymorphic isoenzymes in Pitch pine (*Pinus rigida* Mill.). *Heredity* 40: 27-32.

Hamrick, J.L.; Linhart, Y.B. and Mitton, J.B. (1979) Relationships between life history characteristics and electrophoretically detected genetic variation in plants. *Annu. Rev. Syst. Ecol.* 10: 173-200.

Hamrick, J.L.; Godt, M.J.W. and Sherman-Broyles, S.L. (1992) Factors influencing levels of genetic diversity in woody plant species. *New For.* 6: 95-124.

Hamrick, J.L. and Godt, M.J. W. (1989) Allozyme Diversity in Plants. In: *Plant Population Genetics, Breeding and Genetic Resources* (eds. Brown et al.) Sinauer Associates, Sunderland, M.A., pp. 43-63.

Hertel, H. and Maurer, W.D. (1998) Biochemical-Genetic investigations in Scots pine (*Pinus sylvestris* L). A practical guide to separation methods and zymogramm evaluation.

Hertel, H. (1997) Biochemisch-Genetische Untersuchungen bei Kiefer (*Pinus sylvestris* L). Mitteilungen der Bundesforschungsanstalt fur Forst- und Holzwirtschaft. No. 186.

Hussain, A. (1995) Untersuchungen zur genetischen kontrolle von Isoenzym-Polymorphismen und zur genetischen Struktur von *Pinus roxburghii* Sarg. Dissertation des Doktorgrades des Forstwissenschaftlichen Fachbereichs der Georg-August-Universitat Goettingen.

Khosla, P.K., Sareen, T.S., Sharma, J.C. and Khurana, D.K. (1988) Intra-specific variation in *Pinus roxburghii* Sargent. growth and wood characteristics in needle colour morphotypes. *J. Tree Sci.* 7: 79-84.

Konnert, M. (1995) Isoenzymeuntersuchungen bei Fichte (Picea abies (L.) Karst.) und Weisstanne (*Abies alba* Mill.). Anleitung zur Trennmethodik und Auswertung der Zymogramme. Bayerische Landesanstalt fuer forstliche Saat und Pflanzenzucht, Forstamtsplatz 1, D-83317 Teisendorf, Deutschland.

Ledig, F.T. (1986) Heterozygosity, heterosis and fitness in outbreeding plants. In: Conservation biology: the science of scarcity and diversity (ed. Soule, M.E.) Sinauer Associates, Sunderland, Mass, pp. 77-104.

Loveless, M.D. and Hamrick, J.L. (1984) Ecological determinants of genetic structure in plant populations. *Annu. Rev. Ecol. Syst.* 15: 65-95.

Sagwal, S.S. (1978) *Genetic Variation and Selection of Plus Trees in Chir (Pinus roxburghii) of Himacahl Pradesh.* M.Sc. Thesis, HPU, Agri. Complex, Solan (H.P.).

Sehgal, R.N., Chauhan, K., Dhall, S.P. and Khosla, P.K. (1987) Comparison of progeny characters of trees selected for growth and resin yield in *Pinus roxburghii* Sarg. *J. Tree Sci.* 6: 4-6.

Sharma, K.; and G. Von Wuehlisch.(1998) Genetic interpretation of Malate Dehydrogenase (MDH) isozyme gene loci using a new staining approach and the genetic control of ten other isozymes in *Pinus roxburghii* Sarg. *Silva. Genet.* 47: 5-6: 321-332.

Sharma, K.; Degen, B.; Wuehlisch, G. Von and Singh, N.B. (2002) Allozyme variation in eight natural populations of *Pinus roxburghii* Sarg. in India. *Silva. Genet.* 51: 246-253.

Shaw, C.R. and Prasad, R. (1970) Starch gel electrophoresis of enzymes: a compilation of recipes. *Biochem. Genet.* 4: 297-320.

Thormann, R. And Stephan, B.R. (1993) Interpretation of isozyme patterns of malate dehydrogenase in Scots pine using two different staining methods. *Silva. Genet.* 42: 5-8.

Uniyal, D.P. and Rawat, M.S. (1988) Tree Improvement in Uttar Pradesh Hills. North-Western Regional Conference of Silviculturists and Research Workers, Nanital Vakshshya, R.K.

Yeh, F.C. and El-Kassaby, Y.A. (1980) Enzyme variation in natural populations of Sitka spruce (*Picea sitchensis* (Bong.) Carr.). I. Genetic variation patterns among trees from ten IUFRO Populations. *Can. J. For. Res.* 10: 415-422.

Forest Biotechnology in India © 2008
Editors: S.A. Ansari, C. Narayanan & A.K. Mandal, pp.137-148
Satish Serial Publishing House, Delhi

13

A Comparative Phenetic Studies in Casuarinas Using Morphometric and ISSR Markers

R. KAMALAKANNAN, S. BARTHWAL
P. CHEZHIAN, T. BALASARAVANAN, R. YASODHA
K. GURUMURTHI AND M. GHOSH

Division of Plant Biotechnology, Institute of Forest Genetics and Tree Breeding, Coimbatore, 641 002, India

ABSTRACT

Morphological and genetic diversity among three species of *Casuarina* (*C. equisetifolia, C. glauca* and *C. junghuhniana*) and two species of *Allocasuarina* (*A. huegeliana* and *A. littoralis*) were studied with 18 morphometric parameters and six ISSR primers. The morphometric parameters, viz., plant height, branch length, internode length and teeth length differentiated most of the species and revealed a significant diversity. ISSR-PCR analysis was conducted to estimate the genetic relationship among and within the species. A total of 241 polymorphic bands were amplified in 46 individuals belonging to five species with six ISSR primers. Amplicon size ranged from 220 to 1710 bp. The dendrogram based on ISSR and morphometric markers revealed a more close relationship between *C. junghuhniana* and *A. huegeliana* (genetic distance = 0.887) than that between *A. littoralis* and *A. huegeliana* (genetic distance = 0.983).

Keywords : *Allocasuarina, Casuarinaceae,* genetic diversity, ISSR, morphometry.

INTRODUCTION

Family Casuarinaceae comprises four genera (*Allocasuarina, Casuarina, Ceuthostoma, Gymnostoma*) and 97 species (Chen, 1982: Johnson and Wilson, 1989). *Casuarina* extends to warm temperate regions and *Allocasuarina* in warm to cool temperate regions of Australia (Steane *et al.*, 2003). Among the four genera, the species of *Casuarina* L. Johnson and *Allocasuarina* L. Johnson are commercially cultivated in many tropical and subtropical regions of the world, while the other two genera *Gymnostoma* L. Johnson and *Ceuthostoma* L. Johnson occur in wild. *Casuarina* spp. were introduced to India as a fuelwood species during the late 19th century (Nicodemus *et al.*, 2001) and used for landscaping, pulp, lumber, medicine, tannin, dye and sand-shifting control in coastal areas (Pan *et al.*, 1996). Considerable difference in growth and morphological traits among *Casuarinas* spp. has been reported from different regions. Steane *et al.* (2003) reported extreme morphological reduction in this family, as well as the unique combination of morphological traits. *Allocasuarina* differs from *Casuarina* in terms of leaves on branchlets. In *Allocasuarina,* cone bracteoles are thick and woody, extending slightly beyond cone body, with a separate angular or divided abaxial protuberance; mature samaras are red-brown to black and shiny (Chen, 1982).

Taxonomic and phylogenetic relationships within the species of casuarinas have been studied using chromosome number, pattern of geographical distribution (Barlow, 1959, 1983) and diversification in the morphological characteristics (Wilson and Johnson, 1989; Kumar and Gurumurthi, 2000), DNA markers (Ho *et al.*, 2002; a,b, 2004; Yasodha *et al.*, 2004) and plastid *matK* and *rbcL* (Sogo *et al.*, 2001). Analysis of the genetic variation within and among populations of a species will provide the basic information for breeding programmes. The present study was conducted to analyse the genetic relationship within and among populations of three *Casuarina* and two *Allocasuarina* species using morphological and ISSR markers.

MATERIALS AND METHODS

Plant materials

Seeds of three *Casuarina* species (*C. junghuhniana* Miq., *C. glauca* Sieb ex Spreng and *C. equisetifolia* L) and two *Allocasuarina* species (*A. huegeliana* (Miq.)L.Johnson and *A. littoralis* (Salisb.) L. Johnson) were obtained from Australian Tree Seed Center, CSIRO, Australia (Table 1). The seedlings were raised in the germplasm bank of Institute of Forest Genetics and Tree Breeding, Coimbatore, India and were further used for raising a species trial at Panampally, Kerala, India. Juvenile shoot tips (needle) were used for DNA extraction.

Table 1: Details of seed sources of *Casuarina* and *Allocasuarina* species used for morphological and molecular study.

S.No	Species	Gametic Number (n)	CSIRO Seed lot no.	Origin
1	*Allocasuarina huegeliana*	13	15801	Sanford Rock (WA)
2	*Allocasuarina littoralis*	11	13876	GordanandChili, Cks (QLD)
3	*Casuarina equisetifolia*	9	19129	Lakei/sibur Bako (MALAYSIA)
4	*Casuarina glauca* Sieb Ex Spreng	9	15941	Burrum Heads (QLD)
5	*Casuarina junghuhniana*	9	19489	Kapan Kupang (IN DONESIA)

Morphological characterization

Eighteen morphological characters (seven qualitative and eleven quantitative) were assessed from 46 individuals of each species (Table 2). Straight stem, pyramidal crown, upright branch, self-pruning capacity, smooth bark and absence of disease were considered as positive qualitative traits. For each quantitative trait, the mean value of all the 46 individuals were calculated and fixed as the standard point. Values lower than the standard point with respect to number of permanent branches, branch thickness, number of nodes, number of teeth, teeth length and internodal length, were considered as positive traits; whereas higher values were considered as negative traits. Values higher than the standard point in terms of plant height, collar diameter and girth at breast height were considered as positive traits, while lower values were considered as negative traits. Positive trait was scored as "1" and the negative trait as "0".

Molecular diversity analysis

Genomic DNA was isolated from 100 mg juvenile needles collected from 25 individuals of each species by using QIAGEN DNeasy plant

Table 2: List of morphological characters used to study genetic relationship among five *Casuarina* species.

Morphological Character	Average	Scored as "0"	Scored as "1"
Quantitative			
1. Plant height (m)	2.92	<Average	>Average
2. Collar diameter (mm)	36.29	<Average	>Average
3. Girth at Breast Height - GBH (mm)	14.36	<Average	>Average
4. Number of permanent branches	18.48	>Average	<Average
5. Branch length (cm)	135.34	>Average	<Average
6. Needle length (cm)	34.73	>Average	<Average
7. Thickness of needle (mm)	1.04	>Average	<Average
8. Number of nodes per needle	50.82	>Average	<Average
9. Number of teeth per node	12.83	>Average	<Average
10.Teeth length (mm)	2.51	>Average	<Average
11. Internodal length (cm)	0.86	>Average	<Average
Qualitative			
1. Stem form	-	Bent	Straight
2. Crown shape	-	Open	Pyramid
3. Branch Angle		>60°	<60°
4. Self-pruning	-	Absent	Present
5. Bark surface	-	Rough	Smooth
6. Bark ring	-	Present	Absent
7. Disease attack	-	Present	Absent

mini kit (Hilden, Germany) following manufacturer instructions. Extracted DNA was quantified using spectrophotometer and by comparing band intensities with known standards of lambda DNA (Bangalore Genei, India) on 0.8% agarose gels. A working solution of DNA (30 ng/µl) was prepared in sterile double distilled water. The PCR mixture (10 µl reaction) contained 1.0 µl 10 × PCR buffer, 2.5 mM $MgCl_2$, 40 µM dNTP mix, 0.3 U Taq DNA polymerase (Bangalore Genei, India), 100 nM ISSR primer (University of British Colombia, Canada and Centre for DNA Fingerprinting and Diagnostics, Hyderabad, India) and 15 ng of template DNA. Six ISSR primers were used in the present study (Table 3). The amplification reactions were carried out in Peltier Thermal Cycler (PTC-200; MJ Research, USA) using the following program: 3 min at 94°C for initial denaturation; 35 cycles of 30 s at 94°C for denaturation, 30 s at 50°C for annealing and 1 min at 72°C for extension; 72°C for 10 min for final extension. The amplification products were separated on 2% agarose gel, stained with ethidium bromide and visualized under uv light.

Table 3: List of ISSR primers used to study genetic relationship among the individuals of *Casuarina* and *Allocasuarina* species.

Primer code	Nucleotide sequence
5′ anchored	
$R(CA)_7$	5′ GRTRCYGRTRCACACACACACACA 3′
$T(GT)_9$	5′ CRTAYGTGTGTGTGTGTGTGTGT3′
$TY(CAG)_4$	5′ ARRTYCAGCAGCAGCAG 3′
3′ anchored	
$(GA)_8R$	5′ GAGAGAGAGAGAGAGARGY 3′
UBC810	5′ GAGAGAGAGAGAGAGAT 3′
UBC842	5′ GAGAGAGAGAGAGAGAYG 3′

Data analysis

The binary data, scored from morphological observations and ISSR profiles, were analysed using NTSYSpc Software (Rohlf, 1992). Similarity matrix, based on Nei and Li's similarity indices (Nei and Li, 1979) was used for UPGMA cluster analysis. Percent polymorphism, allele frequency (Kimura and Crow, 1964), Shannon's information index (Lewontin, 1972) and genetic identity/diversity (Nei, 1972) within and between the five species, were analysed assuming Hardy-Weinberg equilibrium and using software, POPGENE 1.32 (Yeh *et al.*, 1997). Hierarchical analysis of molecular variance (AMOVA; Excoffier *et al.*, 1992) was performed as described by Huff *et al.*, 1993 using ARLEQUIN software (V. 2.0). For AMOVA the populations were divided into two groups as *Casuarina* and *Allocasuarina* species and for all analysis the number of permutations for testing significance was set as 1023.

RESULTS

Morphological analysis

All the eighteen morphological traits showed significant differences among the five species (Table 4). Among the different species, maximum genetic diversity in terms of percent polymorphism was observed in *C. junghuhniana* (50%) followed by *C. equisetifolia* (44.4%) and *A. littoralis* (38.9%), whereas minimum genetic diversity was recorded in *A. heugeliana and C. glauca* (33.3%).

Table 4: Basic statistics for eleven morphological parameters analysed in five *Casuarina* species.

Character	*C.equisetifolia*		*C. glauca*		*C. junghuhniana*		*A. heugliana*		*A. littoralis*	
	Mean	SD	Mean	SD	Mean	SD	Mean	SD	Mean	SD
Plant height (m)	3.6	0.6	1.7	0.4	2.9	0.5	1.9	0.3	3.6	0.4
Collar diameter (mm)	46.5	9.5	18.5	4.6	39	6	29.6	6.1	40.2	4.6
Girth at Breast Height (mm)	22.9	6.5	4.5	1.2	13.6	6.5	4.2	1.6	17.6	3.3
No.of permanent Branches	28.2	6.3	6	2.3	17.9	9.3	12.4	4.6	16.2	6.5
Length of Branch (cm)	131.2	28.2	117.7	38.3	144	33.1	115.9	24.9	110.9	44.2
Length of Needle (cm)	25.7	3.7	37.1	13	44.2	7.9	45.5	7.8	32.3	4.1
Thickness of Needle (cm)	0.9	0.1	1.2	0.1	1.2	0.1	1.3	0.1	0.8	0.1
No. Nodes/Needle	49.3	6.3	61	13.8	60	7.4	45.4	6.4	57.2	6
Teeth/Node	12	0	15	0	15	0	16	0	10	0
Teeth length (mm)	1.5	0.1	2	0.2	4	0	3.3	0.5	2	0
Internode length (cm)	0.7	0.1	0.8	0.2	0.9	0.1	1.2	0.2	0.7	0.1

SD=Standard Deviation

Nei's genetic identities and distances between the three *Casuarina* and two *Allocasuarina* species populations are shown in Table 5. Maximum similarity was observed between *C. junghuhniana* and *A. heugeliana* (0.887) whereas maximum genetic distance was found between *A. littoralis* and *A. huegeliana* (0.983) as depicted in the dendrogram (Fig. 1).

Table 5: Nei's measures of genetic similarity (above diagonal) and genetic distance (below diagonal) between the species of *Casuariuna* and *Allocasuarina*.

	C. equisetifolia	*C. glauca*	*A. heugeliana*	*C. junghuhniana*	*A. littoralis*
C. equisetifolia	-	0.581	0.559	0.650	0.704
C. glauca	0.542	-	0.750	0.768	0.576
A. heugeliana	0.582	0.288	-	0.887	0.374
C. junghuhniana	0.430	0.264	0.120	-	0.475
A. littoralis	0.351	0.552	0.983	0.745	-

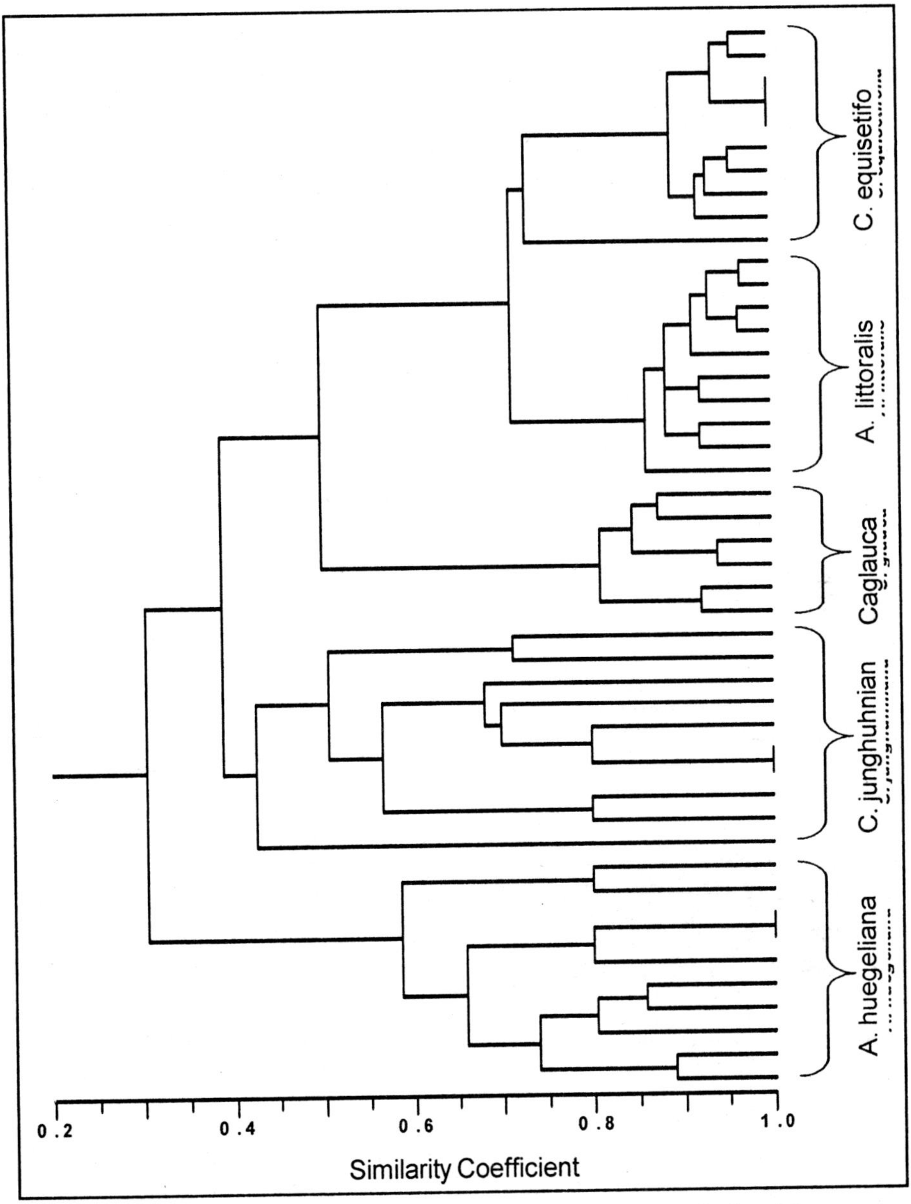

Fig. 1. Dendrogram based on morphomertic data showing relationships among different species of *casuarinas*.

A wide variation was observed in all the five species which differed from each other in one or more individual characters. This confirmed that all the species were morphologically distinct as supported by UPGMA dendrogram (Fig. 1).

Molecular analysis

The six ISSR primers amplified 241 loci of fragment size ranging from 220 to 1710 bp in two *Allocasuarina* and three *Casuarina* species (Fig. 2). Among the five species studied, maximum genetic diversity in terms of per cent polymorphic loci was observed in *C. equisetifolia* (18.55%), while the minimum was in *A. littoralis* (4.84%). Maximum similarity index (0.759) was observed between the species *C. equisetifolia* and *C. glauca*, whereas genetic distance was found to be high (0.660) between *A. littoralis* and *A. huegeliana*. A considerable genetic relatedness was also found between the populations of *C. junghuhniana* and *A. huegeliana* (Table 6).

Table 6: Nei's measures of genetic similarity (above diagonal) and genetic distance (below diagonal) between the species populations of *Casuariuna and Allocasuarina* using ISSR markers.

	C. equisetifolia	*C. glauca*	*A. heuegeliana*	*C. junghuhniana*	*A. littoralis*
C. equisetifolia	-	**0.7585**	0.6577	0.6271	0.6306
C. glauca	0.2764	-	0.5471	0.5230	0.5809
A. heugeliana	0.4190	0.6032	-	0.5873	0.5168
C. junghuhniana	0.4666	0.6481	0.5323	-	0.5522
A. littoralis	0.4611	0.5431	**0.6601**	0.5938	-

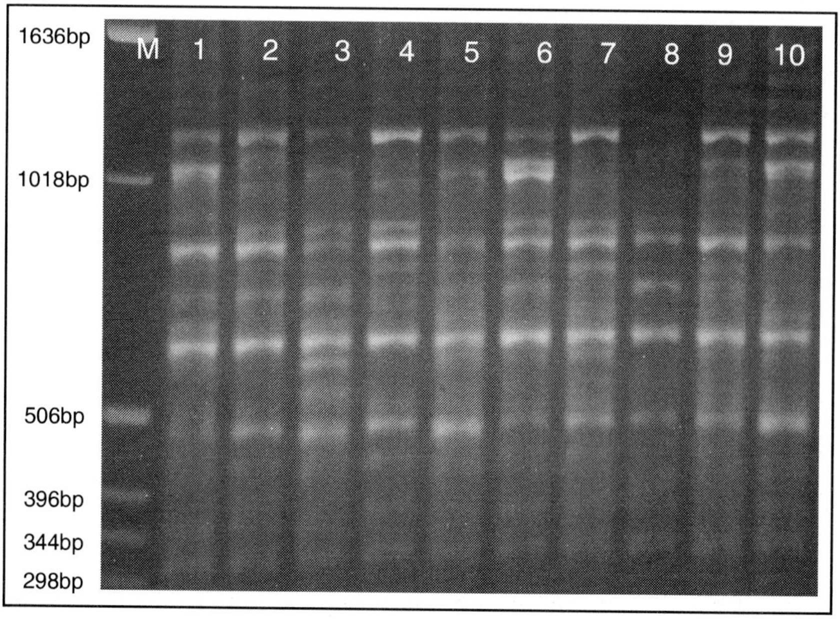

Fig. 2. ISSR profile of *Casuarina equisetifolia* using primer UBC 810.

The UPGMA dendrogram constructed by cluster analyses using similarity matrix for three *Casuarina* and two *Allocasuarina* species showed narrow genetic diversity among the individuals. *A. littoralis*. *A. huegeliana* and *A. littoralis* were separated in two groups. *C. equisetifolia* and *C. glauca* were grouped together but *C. junghuhniana* grouped with *A. huegeliana* (Fig. 3). Results of AMOVA revealed variation among and within species were 92.36 ($p = 0.11$) and 8.11 ($p < 0.38$) respectively (Data not shown) among groups. Variation components among groups was negative (-0.47; $p > 0.26$).

DISCUSSION

An attempt was made to assess genetic relationships and variability in *Casuarina* and *Allocasuarina* species using morphological and molecular markers. The levels of genetic variability estimated using morphometric parameters were high when compared to estimates obtained using ISSR markers. In contrast, there was agreement between UPGMA dendrograms derived from morphological and molecular data. In a similar study, dendrograms obtained from isozyme and RAPD data in *Acacia* species were consistent with morphological data (Casiva *et al.*, 2002). In the present study, the most important characters differentiating species were plant height, girth at breast height, number of permanent branches, internodal length, teeth length and branch length. Morphological analysis clearly separated *Casuarina* and *Allocasuarina* species into two groups. But in the ISSR dendrogram using ISSR markers, *Allocasuarina* grouped with *Casuairna* confirming our earliar report (Yasodha *et al.*,2004). It is noteworthy that *A. huegeliana* is morphologically similar to *C. junghuhniana* in terms of needle length, needle thickness, and number of teeth per node, length of internodes and teeth length. Based on morphology and *matK* sequence data, Steane *et al.* (2003) also concluded that *Casuarina* and *Allocasuarina* are sister taxa.

To our knowledge, this is the first report on assessment of genetic relationships and variability in *Casuarina* species using ISSR and morphological markers. Morphological and ISSR markers were successfully used to detect very high levels of genetic diversity to differentiate *Casuarina* and *Allocasuarina*.

ACKNOWLEDGEMENT

The authors thank Department of Biotechnology, Ministry of Science and Technology, Govt. of India, for financial assistance.

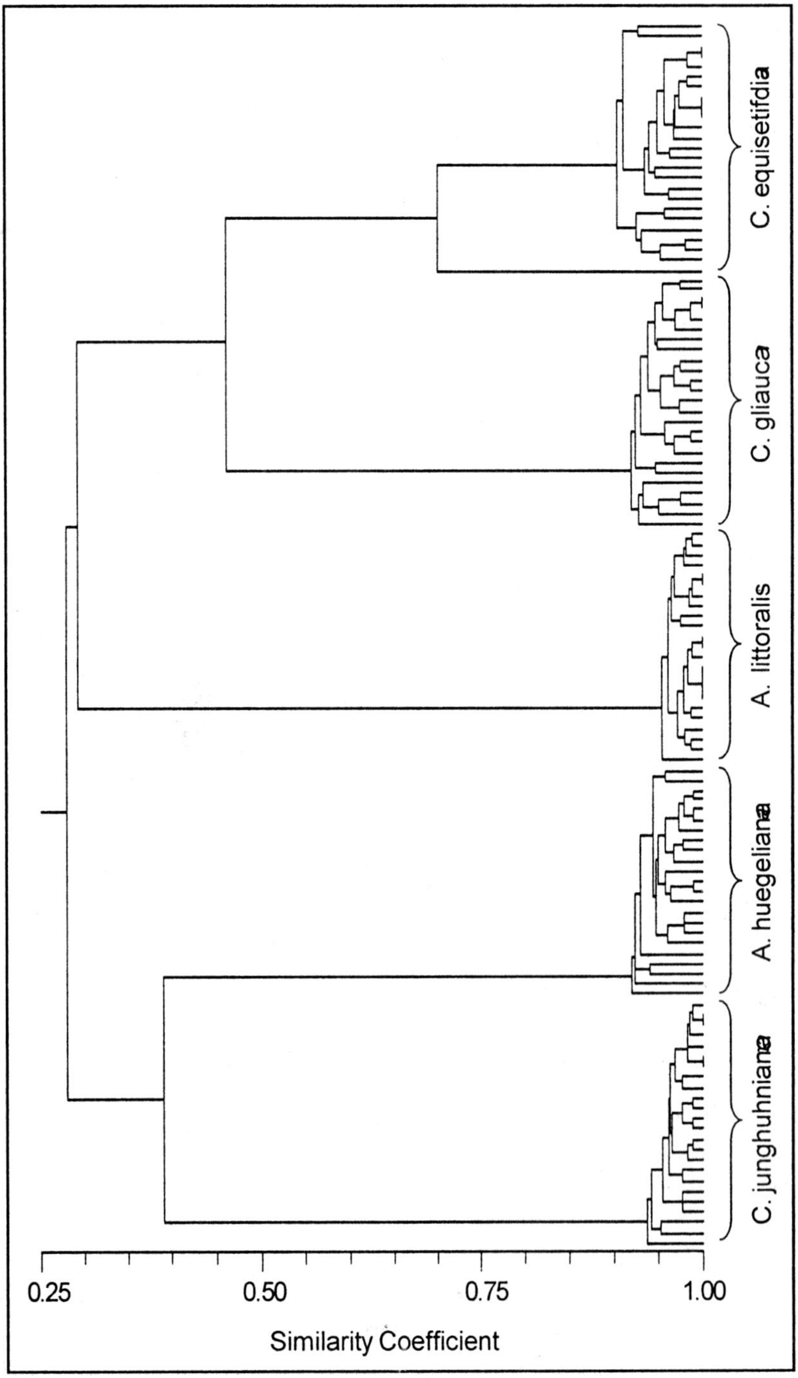

Fig. 3. UPGMA clustering of *Casuarina* spp. and *Allocasuarina* spp. based on Jaccard's similarity coefficient.

References

Barlow, B.A. (1959) Chromosome numbers in Casuarinaceae. *Aus. J. Bot.* 7: 230-237.

Barlow, B.A. (1983) The *Casuarinas*: A Taxonomic and Biogeographic Review. In: *Casuarina Ecology, Management and Utilization* (eds. Midgley S.J., Turnbull J. and Jhonson, R.D.) CSIRO, Australia, pp. 10-18.

Casiva, P.V., Saidman, B.O., Vilardia, J.C. and Ciadldella M. (2002) First comparative phenetic studies of Argentinean species of Acacia (Fabaceae), using morphometric, Isozymal, and RAPD approaches. *Amer. J. Bot.* 89: 843-853.

Chen, T. (1982) Casuarinaceae. In: *Flora* (ed. Tseng Yung-chien) *Reipublicae Popularis Sinicae* 20: 1-5.

Doran, J. and Hall, N. (1983) Notes on Fifteen Australian Casuarina Species. In: *Casuarina Ecology, Management and Utilization* (eds. Midgley S.J., Turnbull, J. and Jhonson, R.D.) CSIRO, Australia, pp 19-52.

Excoffier, L., Smouse, P.E. and Quattro, J. M. (1992) Analysis of molecular variance inferred from matrix distances among DNA haplotypes: application to human mitochondria DNA restriction data. *Genetics* 131 : 479-491.

Ho. K.Y., Ou, C.H., Yang, J.C. and Hsiao, J.Y. (2002 a) An assessment of DNA polymorphisms and genetic relationships of *Casuarina equisetifolia* using RAPD markers. Bot. *Bull. Acad. Sin.* 43: 93-98.

Ho, K.Y., Yang, J.C. and Hsiao, J.Y. (2002 b) An assessment of genetic diversity and documentation of hybridization of *Casuarina* grown in Taiwan using RAPD markers. *Int. J. Plant Sci.* 163: 831-836.

Ho, K.Y., Yang, J.C., Deng, S.L. and Chen, T.H. (2004) Assessment of genetic variation and relationships of international provenances of *Casuarina junghuhniana* using ISSR. *Taiwan J. For. Sci.* 19: 79-88.

Huff, D.T., Peakall T. and Smouse. P.E. (1993) RAPD variation within and among natural population of outcrossing buffalograss (*Buchlo dactyloides* (Nutt.) Engelm). *Theor. Appl. Genet.* 86: 927-934.

Johnson, L.A.S. and Wilson, K.L. (1989) Casuarinaceae: a synopsis. In: *Evolution, Systematics, and Fossil History of the Hamamelidae, Volume 2: Higher Hamamelidae* (eds. Crane, P.R. and Blackmore, S.) Systematics Association Special Volume No. 40B, Clarendon Press, Oxford, pp. 167-188.

Kimura, M. and Crow, J. (1964) The number of alleles that can be maintained in a finite population. *Genetics* 49: 725-738.

Kumar, A. and Gurumurthi, K. (2000) Genetic divergence studies on clonal performance of *Casuarina equisetifolia*. *Silva. Genet.* 49: 57-60.

Lewontin, R.C. (1972) The Apportionment of Human Diversity. *Evol. Biol.* 6 : 381-398.

Nei, M. (1972) Genetic distance between populations. *Am. Natur.* 106: 283-292.

Nei, M and Li, W.H. (1979) Mathematical models for studying genetic variation in terms of restriction endonucleases. *Proc. Nat. Acad. Sci. USA*. 76: 5269-5273.

Nicodemus, A., Varghese, M. and Nagarajan, B. (2001) Genetic Improvement of *Casuarina equisetifolia* through Selection and Breeding, In: *Casuarina: Improvement and Utilization* (eds. Gurumurthi, K., Nicodemus, A. and Siddappa). Institute of Forest Genetics and Tree Breeding, Coimbatore, India, pp.3-14.

Pan, Y., Li, Y. and Tan, T.Y. (1996) Casuarina provenance test. *For. Res.* 9: 138-145.

Rohlf, F.J. (1992) NTSYS-pc.Numerical taxonomy and multivariate analysis system, Version 1.80, *Applied Biostatistics Inc*, New York.

Sogo A, Setoguchi, H., Noguchi, J., Jaffre, T. and Tobe, H. (2001) Molecular phylogeny of based on *rbcl* and *matK* gene sequences. *J. Plant Res.* 114: 259-464.

Steane, D.A., Wilson, K.I. and Hill, R.S. (2003) Using *matK* sequence data to unravel the phylogeny of Casuarinaceae. *Mol. Phylogenet. Evol* .28: 47-59.

Wilson, K. L and Johnson, L.A.S. (1989) *Casuarinaceae. Flora of Australia* 3: 100-174.

Yasodha, R., Kathirvel, M., Sumathi, R., Gurumurthi, K., Archak, S. and Nagaraju, J. (2004) Genetic analyses of casuarinas using ISSR and FISSR markers. *Genetica* 122: 161-172.

Yeh, F. C., Yang, R.C., Boyle, T.J.B., Ye, Z.H. and Mao, J.X. (1997) Popgene, the user-friendly shareware for population genetic analysis. Edmonton, *Molecular Biology and Biotechnology Centre*, University of Alberta, Canada.

❑❑❑

Forest Biotechnology in India © 2008
Editors: S.A. Ansari, C. Narayanan & A.K. Mandal, pp.149-155
Satish Serial Publishing House, Delhi

14

Repetitive Somatic Embryogenesis from Zygotic Embryos of Teak (*Tectona grandis* L.) Cultured *in vitro*

K.R. KRISHNADAS AND E.M. MURALIDHARAN

Biotechnology Division, Kerala Forest Research Institute
Peechi, Thrissur 680653, Kerala, India

ABSTRACT

Somatic embryogenesis and repetitive embryogenesis were observed in isolated immature zygotic embryos cultured on a solid medium consisting of minerals and vitamins of the Woody Plant medium (WPM) supplemented with sucrose 40 g^{-1} and Picloram (2 to 8 mg^{-1}) at different concentrations either singly or in combination with BAP (0.5, 1.5 and 2.0 mg^{-1}). Repetitive somatic embryogenesis was also observed in some of the embryos on transfer to fresh medium. The low frequency maturation of somatic embryos was observed, albeit, development of plantlets could not achieved.

Keywords : *Tectona grandis,* somatic embryogenesis, repetitive embryogenesis, maturation

INTRODUCTION

Teak (*Tectona grandis* L) is one of the most valuable timbers of the world due to its durability, strength, workability and attractiveness of the wood. Teak is also one of the most widely planted hardwood tree species and its cultivation is being undertaken in several countries outside its natural range. Genetic improvement through conventional tree breeding and genetic engineering (Nair, 2004; Dhakate *et al.*, 2005) is expected to improve productivity of cultivated teak.

Micropropagation of teak from seeds, nodes and shoot tips of adult trees of teak has been reported by several workers (Gupta *et al.*, 1980; Mascarenhas *et al.*, 1987; Devi, 1994; Monteuuis, 1994; Tiwari *et al.*, 1995, 2002; Bonal and Monteuuis, 1997). Superior clones of teak can be multiplied on commercial basis through enhanced axillary bud proliferation and rooting. However, the method is not amenable for studies on genetic transformation and somaclonal variation, where regeneration of plantlets from individual cells and tissues are required. Somatic embryogenesis has many advantages over the other methods of micropropagation (Jimenez, 2001). The multiplication rates are typically much higher and the method is amenable to scaling up and automation. Since a separate rooting step is avoided, plantlet regeneration is achieved within a short period. Somatic embryos mostly have a single cell origin and are therefore ideal for genetic transformation and somatic hybridization. Efforts have been made to induce direct and indirect organogenesis and somatic embryogenesis in teak (Kusalkar and Sharon, 1996; Widiyanto *et al.*, 2001; Capote *et al.*, 2002; Daquinta, 2002; Yunaini and Widiyanto, 2003)

Leaves were used as explants by Yunaini and Widiyanto (2003) and shoot tips and leaves by Kusalkar and Sharon (1996) for induction of somatic embryogenesis. Early stages of embryogenesis have been described in these reports but plantlet regeneration was not obtained. In this paper, we report for the first time the induction of somatic embryogenesis in cultures of immature zygotic embryos of teak and the occurrence of repetitive (recurrent secondary) embryogenesis.

MATERIALS AND METHODS

Immature fruits of teak of 0.5 to 1.0 cm dia. were collected from selected trees in the KFRI campus at Peechi, Kerala. After removal of the calyx and the soft downy mesocarp from the endocarp the fruits were washed with dilute detergent, rinsed with distilled water and surface sterilized with 0.1 % $HgCl_2$ for 20 minutes under sterile conditions. The fruits were cut open with a scalpel to extract the immature seeds /embryos. The seeds or embryos were inoculated directly on a solid medium consisting of the minerals and vitamins of WPM (Lloyd and McCown, 1981) supplemented with 40 g^{-1} sucrose and different concentrations of plant growth regulators. In the preliminary experiments, BAP, Kinetin, 2iP, 2,4-D, NAA, Dicamba, TDZ and Picloram were tested for effect on inducing somatic embryogenesis and Picloram was identified as most promising. Further experiments were carried out with WPM and Picloram in the range of 2 to 8 mg-l either singly or along with 0.5 to 2 mg^{-1} BAP. Cultures were incubated in the dark at 25°C ± 2°C. Explants that showed positive response were subcultured to fresh medium every 2-3 weeks. Explants showing secondary embryogenesis were split before subculture. Maturation of somatic embryos was attempted by transfer to WPM supplemented with ABA (0.02 mg^{-1}) and incubation under light (25 mE s^{-1} m^{-2}).

RESULTS

Zygotic embryos at the later stages of development precociously developed into seedlings on the WPM media with different levels of auxins. Upto 50 % of the zygotic embryo explants at the heart shaped stage or globular stage of development gave positive response in terms of showing signs of somatic embryogenesis. Globular pearly white outgrowths on the zygotic embryos were observed in the initial stages (Fig. 1) after about 2 weeks in culture. As the culture progressed further, other developmental stages were observed (Figs 2,3). Callusing was observed in few cultures but this appeared to be originating from tissues other than the zygotic embryo and no further morphogenesis was observed. Better induction of somatic embryos was achieved in media containing picloram along with BAP compared to only picloram. Somatic embryos, inducted and maintained in the media with Picloram, eventually developed callusing and finally turned senescent. Subsequently embryos were transferred to media with 1 mg^{-1} of Picloram along with 0.5 mg^{-1} of BAP to promote further development. Conversion of embryos to plantlets could be avoided on this media even when supplemented with ABA.

Closer observation under a stereo dissection microscope revealed the organized nature of the somatic embryos. The outer surface of the somatic embryos at later stages, particularly the cotyledonary part showed single celled trichomes in the form of minute protuberances with deep purple pigmentation (Fig. 4).

In some of the cultures with late stages of embryos, repetitive embryogenesis was observed at the base of the embryos, after 4 weeks of culture (Fig. 4). Upto 6-8 embryos were produced by each zygotic embryo after three subcultures. Even after transfer to ABA containing media, no improvement in maturation of embryos or change in morphology was observed.

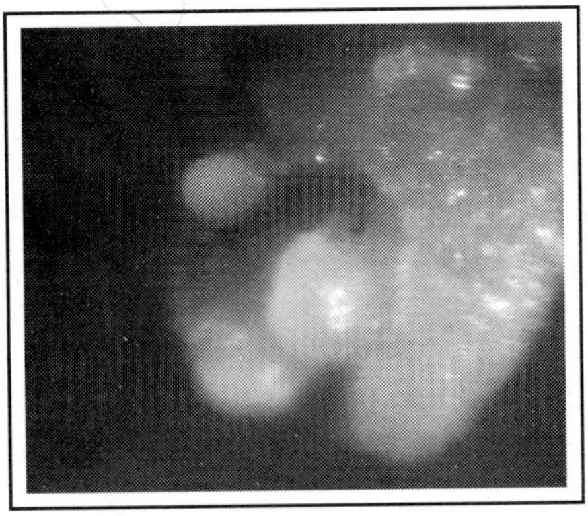

Fig. 1. Induction of somatic embryogenesis in zygotic embryo explants

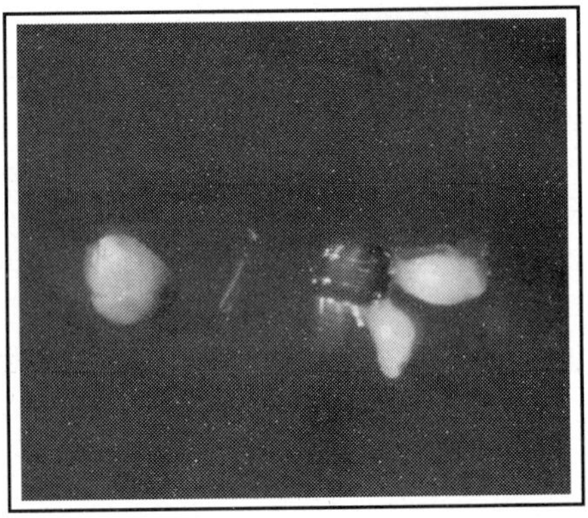

Fig. 2. Somatic embryos at early stages

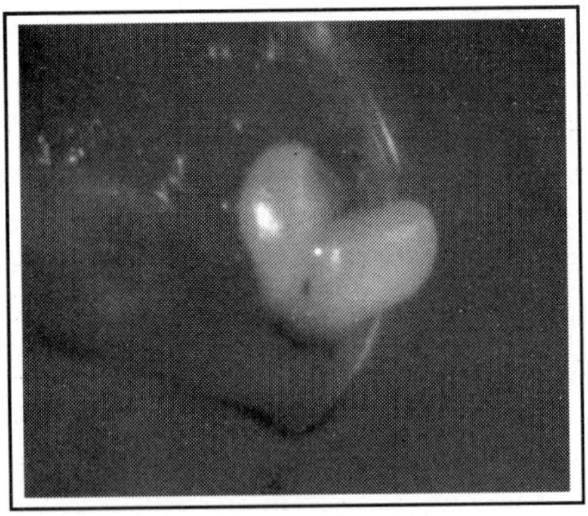

Fig. 3. Somatic embryos at torpedo stage

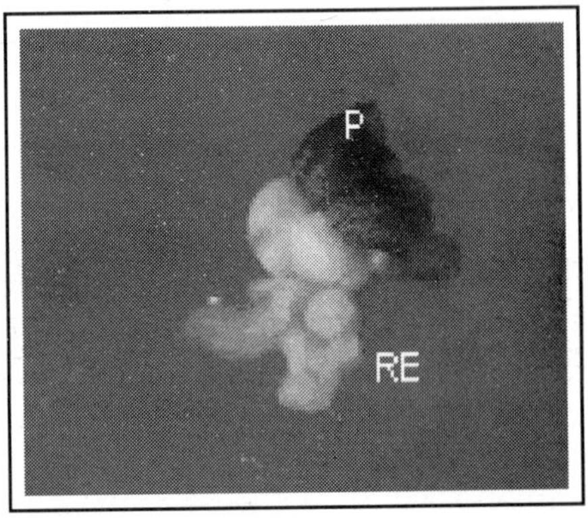

Fig. 4. Repetitive embryogenesis (RE) at base of mature somatic embryo. (P- Pigmentation)

DISCUSSION

Immature zygotic embryos have been widely used as explant for induction of somatic embryogenesis (Mullins and Srivastava, 1976; Muralidharan and Mascarenhas, 1995). The competence of juvenile tissues for somatic embryogenesis has been attributed to the pre-embryonically determined cells in the immature embryos. Auxins are a crucial component of the media used for induction of somatic embryos. However, Picloram has not been commonly used as compared to 2,4-D or NAA, for induction of callus or embryos (Jimenez, 2001).

In most of the reports on somatic embryogenesis, inhibition of somatic embryo development on the induction medium was observed and further development required further transfer to either hormone free medium or medium with lower levels of auxins. In the present study, embryos were transferred to media with low levels of auxin along with BAP to encourage embryo development. The deep purple epidermal trichomes found in the somatic embryos after a few days of culture in this media is similar to the structures observed on zygotic embryos that germinate precociously *in vitro* and also on the lower surface of young leaves and petioles of seedlings. Pigmentation can, therefore, be an indicator for maturation of somatic embryos. No sign of radicle or plumule development was, however, observed in the somatic embryos transferred to hormone free medium. The use of ABA with or without other adjuvants like osmoticum or activated charcoal, for regulating the maturation of somatic embryos, is widely employed in a number of species, including tree species (Jimenez, 2001).

The absence of an intermediate callus stage in somatic embryogenesis is advantageous since it avoids problems due to somaclonal variation. The limitation of low frequency of plantlet regeneration in direct somatic embryogenesis was overcome in this study through repetitive embryogenesis. Repetitive embryogenesis can result in proliferation of the embryogenic culture for several cycles as has been demonstrated in *Eucalyptus* (Muralidharan and Mascarenhas, 1995).

Somatic embryogenesis, particularly in liquid culture, has the potential for scaling up multiplication rate to a level that will be vastly superior to the current micropropagation techniques which involve multiple shoot formation and rooting. Optimization of high frequency embryogenesis, efficient conversion to plantlets and the development of suspension cultures are necessary before the above results can be utilized for large scale clonal propagation. A somatic embryogenic system is also very much amenable to genetic transformation techniques using *Agrobacterium* or biolistics.

CONCLUSION

The potential of immature zygotic embryos for induction of somatic embryogenesis is demonstrated in the study. Although occurring at a lower percentage and restricted to certain developmental stages of the zygotic embryos, the induction of repetitive embryogenesis indicated potential for developing an efficient clonal propagation system. Further experiments are being carried out to induce a high frequency repetitive embryogenic system and suspension cultures.

References

Bonal, D. and Monteuuis, O. (1997) *Ex vitro* survival, rooting and initial development of *in vitro* rooted vs unrooted micro shoots from juvenile and mature *Tectona grandis* genotypes. *Silva. Genet.* 46: 301-306.

Capote, I., Daquinta, M., Escalona, M., Rodríguez, R., Lezcano, Y., Ramos, L. (2002) Morfogénesis *in vitro* de Teca (*Tectona grandis* L.). *Biotecnología Vegetal* 1(1): 137-44.

Daquinta, M. (2002) Induccion de callos y regeneración de plantas de Teca. Rev. *Biotecnología Vegetal* 2 (1): 15-9.

Devi, Y.S., Mukherjee, B.B., Gupta, S. (1994) Rapid cloning of elite teak (*Tectona grandis* Linn.) by in *vitro* multiple shoot production. *Indian J. Expt. Biol.* 32: 668-671.

Dhakate, P.M., Bakade, G.S., Chavan, M.R. and Sasikumar, S. (2004) Teak Biotechnology: Development and Prospects. In: Bhat, K.M. *et al.* (eds.) Quality Timber Products of Teak from Sustainable Forest Management. KFRI and ITTO, Japan, pp. 438-444.

Gupta, P.K., Nadgir A.L., Mascarenhas, A.F., Jagannathan, V. (1980) Tissue culture of forest trees- Clonal mutiplication of *Tectona grandis* (Teak) by tissue culture. *Plant. Sci. Lett.* 17: 259 -268.

Jimenez, V.M. (2001) Regulation of in vitro somatic embryogenesis with emphasis on the role of endogenous hormones. *R. Bras. Fisiol. Veg.* 13 (2): 196-223

Kushalkar, R. and Sharon, M. (1996) Direct and indirect somatic embryogenesis in teak (*Tectona grandis* L.) *Curr. Sci.* 9: 712-715

Lloyd, G. and McCown, B. (1981) Commercially- feasible micropropagation of mountain laurel, *Kalmia latifoliai* by use of shoot tip culture. *Int. Plant. Prop. Soc. Proc.* 30: 421-427.

Mascarenhas, A.F., Kendurkar, S.V., Gupta, P.K., Khuspe, S.S., Agarwal, D.C. (1987) Teak. In: *Cell and Tissue Culture in Forestry. Vol. 3* (eds. Bonga, J.M. and Durzan, D.J.) Martinus Nijhoff Publishers, Dordrecht, pp. 300-315.

Monteuuis, O. (1994) Recent advances in mass clonal propagation of Teak. *Proc. Int. Workshop BIO-REFOR.* Malaysia, pp. 117-121.

Mullins, M.G. and Srivastava, C. (1976) Somatic embryos and plantlets from an ancient clone of the grape vine (C.V. Cabernet - sauvignon) by apomixis *in vitro. J. Exp. Bot.* 27: 1022 - 1030.

Muralidharan, E.M. and Mascarenhas, A.F. (1995) Somatic Embryogenesis in Eucalyptus. In : Somatic Embryogenesis in Woody Plants, Vol. 2 (eds. Mohan Jain, S., Gupta, P.K. and Newton, R.J.) Kluwer Academic Publishers, Netherlands, pp. 23-40.

Nair, K.S.S. (2004) Pest Factor in the Intensification of Teak Cultivation: A Global Assessment. In: *Quality Timber Products of Teak from Sustainable Forest Management* (eds. Bhat, K.M. *et al.*) KFRI and ITTO, Japan, pp. 438-444.

Tiwari, S. K. and Pandey, R. K. (1995) A preliminary observation on *in vitro* propagation of teak (*Tectona grandis*) through tissue culture from excised apical and seedling explants. *J. Trop. For.*11: 46-50.

Tiwari, S.K., Tiwari, K.P. and Siril, E.A. (2002) An improved micropropagation protocol for teak. *Plant Cell Tissue Organ Cult.* 71:1-6.

Widiyanto, S.N., Erytrina, D. and Rahmania, H. (2001) Adventitious Shoot Formation on teak (*Tectona grandis* L.f.) Callus Cultures derived from Internodal Segments. *Abstract in The 2nd International Symp. on Biotechnology of Tropical and Subtropical species*, Taipei.

Yunaini, L and Widiyanto, S.N. (2003) Somatic Embryogenesis from Callus Cultures of Teak (*Tectona grandis* L.f.) derived from Leaf Explants. *Abstract presented at 2003 Congress on in vitro biology.*

❑❑❑

Forest Biotechnology in India © 2008
Editors: S.A. Ansari, C. Narayanan & A.K. Mandal, pp.157-164
Satish Serial Publishing House, Delhi

15

Micropropagation of Adult *Bambusa balcooa* Roxb. Through Multiple Shoot Formation and Somatic Embryogenesis

T.P. SHALI AND E.M. MURALIDHARAN

Biotechnology Division, Kerala Forest Research Institute
Peechi, Thrissur, 680653, Kerala, India

ABSTRACT

Plantlet regeneration through enhanced axillary bud proliferation, multiple shoot formation and rooting was obtained through *in vitro* culture of nodes collected from mature field grown clumps of *Bambusa balcooa*. Surface sterilized nodal explants of 2-3 cm taken from lateral branches developed sprouts when placed in a liquid medium consisting of the minerals and vitamins of the Murashige and Skoog's medium (MS) supplemented with benzylaminopurine (BAP) and napthalene acetic acid (NAA) and sucrose 2 %. Multiple shoot formation was observed in MS medium supplemented with different levels of BAP and was optimum at 3 mg^{-1}. Subcultures of 4 weeks duration was best for maintaining multiplying shoot cultures. Excised shoot, either single or in clusters of 3-5 shoots showed upto 80 % rooting *in vitro* on MS medium supplemented with IBA or NAA. Rooted shoots were transferred to vermiculite: sand (1:1) mixture and hardened under intermittent mist for 3 weeks followed by transfer to soil and maintained in nursery. Nodal explants taken from actively growing multiple shoots and cultured on MS supplemented with 3 mg^{-1} of 2, 4-Dichlorophenoxyacetic acid (2,4-D) developed a yellowish white calli which on transfer to a medium with 3 mg^{-1} each of 2,4-D, NAA and BAP, turned embryogenic. Transfer to MS with 0.5 mg/l NAA and 1 mg^{-1} BAP resulted in proliferation and development of somatic embryos. Greening of embryos and development of scutellum and root was observed.

Keywords : *Bambusa balcooa,* multiple shoots, embryogenic callus, rooting

INTRODUCTION

Large-scale cultivation of bamboo is expected to be taken up in India following the popularization efforts undertaken in recent times by the National Mission on Bamboo Applications (NMBA) and other agencies. A priority list of commercially important bamboo has been prepared and a huge demand for planting material is anticipated in the next few years. *Bambusa balcooa* Roxb. is one of the most important bamboo species and reported by shows sporadic flowering but no records of seed formation are known (Banik and Alam, 1987). Like several other species of bamboo, vegetative propagation through rhizome offsets and rooted culm cuttings is the common means of propagation. For a large-scale planting programme, these methods are inadequate since the multiplication rates are relatively low.

Micropropagation has been widely reported in bamboo species from explants taken from both seedling as well as adult plants (Chaturvedi *et al.*, 1993; Hirimburegama and Gamage, (1995); Ramanayake and Yakandawala, (1997). Somatic embryogenesis and plantlet regeneration has also been reported in many species of bamboo (Godbole *et al*, 2002; Woods *et al.*, 1992; Yeh and Chang, 1986). A recent report (Das and Pal, 2005) describes a micropropagation procedure for *B. balcooa* from adult clumps.

In this paper we present results of micropropagation experiments carried out with explants taken from adult clumps of *B. balcooa,* and a complete procedure for plantlet regeneration through multiple shoot formation and the occurrence of somatic embryogenesis in callus derived from nodes of shoot cultures .

MATERIALS AND METHODS

Explants used for this study were collected from lateral branches of culms of adult *Bambusa balcooa* growing in the bambusetum at KFRI, Peechi. After removal of leaf sheaths, the stems were washed in a dilute solution of a detergent, wiped with 70 % ethyl alcohol and 3 cm long nodal segments were excised for culture. A pretreatment was given with a solution containing Bavistin (1 % w/v) and gentamycin (200 mg^{-l}) for 45 minutes followed by surface sterilization with 0.1% w/v mercuric chloride solution for 15 min. Later, under aseptic conditions, explants were washed 3-4 times with sterile distilled water and placed in shoot induction medium in test tubes. A liquid medium (MS1) consisting of the minerals and vitamins (MS) of Murashige and Skoog (1962) medium supplemented with 0.5 mg^{-l} of benzylaminopurine (BAP) and 20 g^{-l} of sucrose was used for inducing sprouts from nodes. Cultures were maintained in glass bottles with polypropylene lids without agitation under fluorescent tubelights (25 mE s^{-1} m^{-2}) and 25°C $\pm$ 2°C.

To test the effect of cytokinins on shoot multiplication, sprouted nodes were transferred to a series of liquid shoot multiplication media (MS2) consisting of minerals and vitamins of MS medium with 2 % sucrose supplemented with BAP at low, medium or high levels (1.75 mg/l, 3 mg^{-l} and 6 mg^{-l} respectively) and napthaleneacetic acid (0.1 mg^{-l}). Subcultures were transferred to fresh media every 4 weeks. Single or clusters of shoots were transferred to solid MS3 medium consisting of minerals and vitamins of MS but devoid of NH_4NO_3 with 2 % sucrose and supplemented with different combinations of the auxins, IBA, NAA and IAA. After a week, the shoots were transferred to liquid MS1 for further growth. Rooted shoots were transferred to a mixture of vermiculite and sand (1:1) in airtight and transparent food containers and placed in a green house for 4 weeks for hardening. Surviving plantlets were transferred to soil in polybags and maintained in the nursery.

Nodal explants of about 1 cm were excised from already established shoot cultures and placed on a callus induction medium (MS4) consisting of minerals and vitamins of MS added with 2% sucrose and 2,4-D (3 mg^{-l}) and solidified with 0.7 % agar. Cultures in Petri dishes were maintained in dark at 25°C $\pm$ 2°C. Explants that developed callus were transferred to fresh media every 4 weeks. To induce somatic embryogenesis, calli were transferred to MS media supplemented with various concentrations of 2,4-D, NAA and BAP and incubated in dark. To induce plantlet formation, somatic embryos were transferred to a conversion medium consisting vitamins and minerals of MS without hormones or with BAP (1 mg^{-l}) and NAA (0.5 mg^{-l}) with or without 0.2 % activated charcoal, and incubated under light (25 mE s^{-1} m^{-2}).

RESULTS

In the nodal explants, sprouting of axillary buds occurred within a week of incubation on MS1 media. Each explant produced 2-8 shoots (Fig. 1) but occasionally up to 20 shoot intials were observed. On prolong retaining the explants in the same medium further growth of the shoots was inhibited. Explants continued to grow and produced multiple shoots (Fig. 2) on transfer to MS2 supplemented with three levels of BAP and also containing NAA. A highly prolific shoot culture was obtained in this study even at lower levels of cytokinin. Shoot cultures generally consisted of short shoot initials as well as elongated shoots with fully open leaves. But at higher levels of hormones, shorter shoots with unopened leaves were also produced. Highest multiplication (>30 shoots per subculture) was achieved with 3 mg^{-1} BAP but better shoot morphology and health of shoot cultures was induced with low cytokinin levels where 20-30 shoots were formed per subculture. The *in vitro* flowering in few shoot cultures in the third and subsequent sub-culturing was occasionally noticed on higher cytokinin levels.

Fig. 1: Sprouting of axillary buds in nodal explants.

Fig. 2: Multiple shoot formation.

Single or cluster of shoots transferred to the rooting medium (MS3 with auxins) initiated roots (Fig. 3) in about 7 days. The best response (80%) in terms of induction and number of roots was obtained on media supplemented with 5 mg^{-1} of IBA and NAA. NAA (5 mg^{-1}) gave the next best results but IBA separately at all concentrations gave poor results. Shoots maintained on the rooting medium turned brown and lost their health rapidly. When rooted shoots were shifted from rooting medium to MS2 the plantlets remained healthy and continued development. Plantlets were hardened under mist for 4 weeks and after 7 weeks new leaves developed following which they were transferred to poly bags.

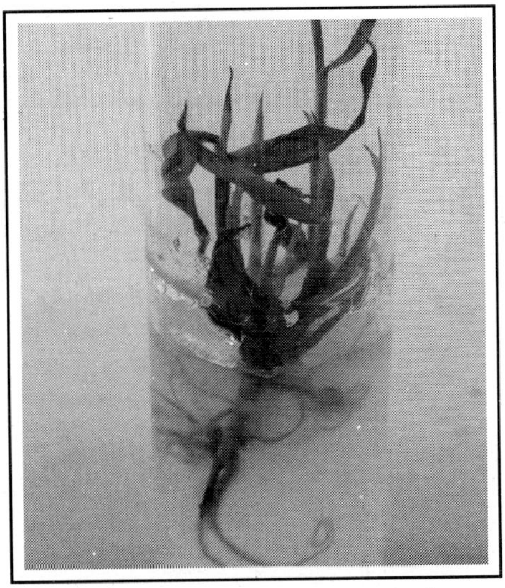

Fig. 3: *In vitro* rooting of shoot clusters.

Nodal explants developed callus at the nodal region (Fig.4) after15 days of culture in MS4. After repeated subcultures on fresh media profuse growth of a yellowish white callus was obtained within two months. On the embryogenesis media with 2,4-D, NAA and BAP (3 mg^{-1} each) a highly embryogenic pearly white callus (Fig. 5) developed. Lower levels of the hormones (1 and 2 mg^{-1} each) resulted in lower levels of embryogenesis. On transfer to a conversion media with lower levels of BAP (1 mg^{-1}) and NAA (0.5 mg^{-1}) but free of 2,4-D green scutella like structures developed in the cultures (Fig. 7) with occasional presence of roots. Calli with mature somatic embryos were also shifted to a hormone free medium with or without activated charcoal, but conversion of embryos to plantlets was not obtained.

DISCUSSION

The use of the liquid medium supported sprouting of axillary buds in nodal explants as well as multiplication of shoot cultures. It also ensured that any contamination especially those of endogenous origin, expressed early in the culture period since the explants were constantly exposed to the medium. On the other hand solid medium greatly slowed the time taken for contamination to express itself and often, endogenous contamination was noticed only after subculture to fresh

Fig. 4: Hardened plantlet.

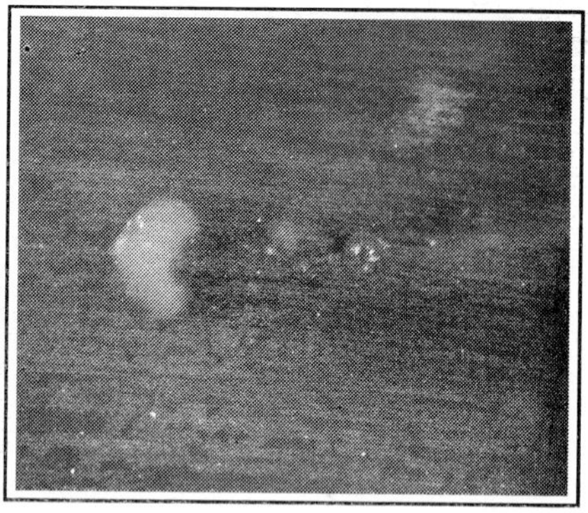

Fig. 5: Induction of callus in nodal explant.

Fig. 6. Highly embryogenic callus.

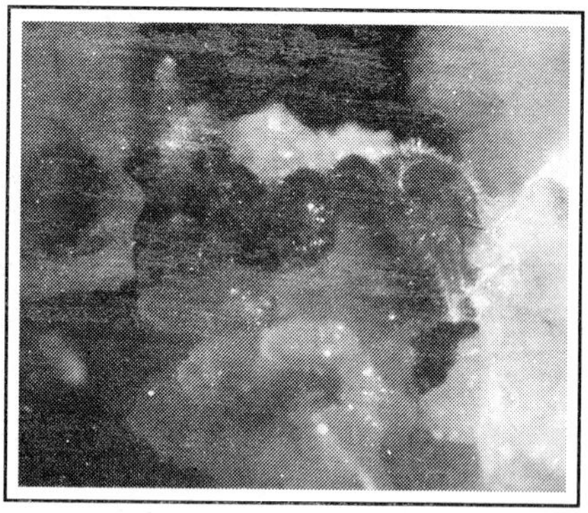

Fig. 7. Mature somatic embryos.

media. The cost reduction obtained with use of liquid media is significant as also the ease of handling during subcultures. It is also possible to further improve the micropropagation procedures through use of semi-automated liquid handling systems.

Multiple shoot formation in the presence of cytokinin has been reported in several species of bamboo (Chaturvedi *et al.*, 1993; Ramanayake and Yakandawala, 1997). BAP, either singly or in addition to kinetin and an auxin, is the commonly used cytokinin in most of the reports. Use of higher levels of BAP did not improve the results. Das and Pal (2005) also used low levels of BAP but with kinetin for multiplication of *B. balcooa* shoots. The occasional *in vitro* flowering in some of the flasks maintained at higher levels of BAP (3 and 6 mg^{-l}) indicates that lower cytokinin levels would be ideal for bamboo provided sufficient multiplication rates are achieved. Ramanayake *et al.* (2001) had also observed *in vitro* flowering in shoot cultures of *D. giganteus* under higher levels of BAP (6 mg/l). High percentage of rooting was obtained in our study when IBA and NAA (5 mg/l each) were used in the rooting media. Das and Pal (2005) obtained the best results in rooting using IBA at low concentrations. In our studies, rooting was much lower with this hormone. Das and Pal (2001) observed seasonal variations in morphogenetic response which may be the governing factor for similar results in our study.

Mukunthakumar *et al.* (1999) reported the formation of callus in nodal explants of *B. vulgaris* grown on media with 3 mg^{-1} 2,4-D and further development of somatic embryos on transfer to media with 2 mg^{-1} BAP and 3 mg^{-1} NAA. Godbole *et al.* (2002) also reported similar results with *Dendrocalamus hamiltonii* where sprouted nodes were used to induce the embryogenic callus. Ramanayake *et al.*(2001) reported organogenesis in callus derived from nodal explants taken from *in vitro* shoots of *D. giganteus.* Use of nodes collected from shoot cultures, not only facilitates use of more number of explants for culture, but also decreases the risk of losing cultures to contamination, which is common during establishment of cultures from field explants.

CONCLUSION

An efficient micropropagation protocol for rapid clonal propagation of *B. balcooa* was achieved through enhanced axillary bud proliferation and rooting of shoots. Use of stationary liquid media for shoot multiplication can reduce cost of propagation. Improvements in rooting percentage and health of the plantlets are needed before the present protocol can be used for a large-scale programme. Further studies on improving the conversion of somatic embryos to plantlets and confirmation of genetic integrity of regenerated plantlets through the use of molecular markers are necessary.

References

Banik, R. L. and Alam, M. K. (1987) A note on the flowering of *Bambusa balcooa Roxb. Bano Biggyan Patrika* 16: 25-29.

Chaturvedi, H. C., Sharma, M. and Sharma, A. K. (1993) *In vitro* regeneration of *Dendrocalamus strictus* Nees through nodal segments taken from field-grown clumps. *Plant Sci.* 91 : 97-101.

Das, M. and Pal, A. (2005) *In vitro* regeneration of *Bambusa balcooa* Roxb.: Factors affecting changes of morphogenetic competence in the axillary buds. *Plant Cell Tissue Organ Cult.* 81: 109-112.

Godbole, S., Sood, A., Thakur, R., Sharma, M., and Ahuja, P.S. (2002) Somatic embryogenesis and its conversion into plantlets in a multipurpose bamboo, *Dendrocalamus hamiltonii* Nees et Arn. Ex Munro. *Curr. Sci.* 83: 885-889

Hirimburegama, K. and Gamage, N. (1995) Propagation of *Bambusa vulgaris* (yellow bamboo) through nodal bud culture. *J. Hortic. Sci.*70: 469-475

Mukunthakumar, S., Mathur, J., Nair, P. K. K., and Mathur, S. N. (1999) *In vitro* morphogenetic responses of nodal explants of golden yellow bamboo - *Bambusa vulgaris* Schrad. *J. Non-Timb. For. Prod.* 6: 39-43

Murashige, T and Skoog, F. (1962) A revised medium for rapid growth and bio assay with tobacco tissue culture. *Physiol. Plant.* 15 : 473-497

Ramanayake, S. M. S. D and Yakandawala, K. (1997) Micropropagation of the giant bamboo (*Dendrocalamus giganteus* Munro) from nodal explants of field grown culms. *Plant Sci. Limerick* 129: 213-223

Ramanayake, S. M. S. D., Wanniarachchi, W. A.V. R., Tenakoon, T. M. A. (2001) Axillary shoot proliferation and in vitro flowering in an adult giant bamboo *Dendrocalamus giganteus* Wall. Ex Munro. *In Vitro Cell. Dev. Biol. Plant* 37: 667-671

Ramanayake, S. M. S. D. and Wanniarachchi, W. A.V. R. (2001) Organogenesis in callus derived from an adult giant bamboo (*Dendrocalamus giganteus* Wall. Ex Munro). *Sci. Hortic.* 98: 195-200

Woods, S. H., Phillips, G. C., Woods, J. E. and Collins, G. B. (1992) Somatic embryogenesis and plant regeneration from zygotic embryo explants in Mexican weeping bamboo, *Otatea acuminata aztecorum*. *Plant Cell Rep.* 11: 257-261

Yeh, M. L. and Chang, W. C. (1986) Somatic embryogenesis and subsequent plant regeneration from inflorescence callus of *Bambusa beecheyana* Munro var. *beecheyana*. *Plant Cell Rep.* 6: 409-411.

□□□

Forest Biotechnology in India © 2008
Editors: S.A. Ansari, C. Narayanan & A.K. Mandal, pp.165-182
Satish Serial Publishing House, Delhi

16

Rapid and Simplified Method of Micropropagation of *Pseudoxytenanthera stocksii*

P.V. SOMASHEKAR, T.S. RATHORE AND K.S. SHASHIDHAR

Tree Improvement and Propagation Division, Institute of Wood Science and Technology, 18th Cross, Malleswaram, Bangalore–560 003

ABSTRACT

This paper describes a rapid and simplified method for mass clonal propagation of *Pseudoxytenanthera stocksii* Munro, a multipurpose solid bamboo which has high demand in agroforestry. Nodal shoot segment obtained from Candidate Plus Clumps exhibited high frequency (100%) multiple shoot induction on the MS liquid medium with additives [ascorbic acid (50 mg/l), citric acid (25 mg/l) and cysteine (25 mg/l)] within 2 weeks. *In vitro* shoots were further multiplied on MS liquid medium with additives and 1.0 mg/l BAP. Sub-culturing was found essential within 2 weeks and multiplication rate increased 5 fold within 4 weeks. Reduced concentration of BAP in the medium favoured shoot multiplication and rooting. Maximum (100%) the *in vitro* rooting was obtained on MS/4 medium with NAA (1.0 mg/l). *In vitro* shoots pulse treated with NAA/ IBA (1000 ppm) exhibited 95% *ex vitro* rooting in sand bed in mist chamber within 3 weeks period. After *ex vitro* rooting, clonal plants with 4-5 tillers and miniature rhizome can be obtained within 4 months period. Following the above protocol, about 5000 plants were produced and used in field trials.

Keywords : *Pseudoxytenanthera stocksii,* micropropagation, Candidate Plus Clump, shoot multiplication, *in vitro* and *ex vitro* rooting.

INTRODUCTION

The genus *Pseudoxytenanthera* is endemic to South India and Srilanka. From South India, four species are reported viz., *Pseudoxytenanthera bourdillonii, P. monadelpha, P. ritcheyi* and *P. stocksii. P. stocksii,* locally known a seeme bamboo, occurs naturally in the Western Ghats. It is distributed in the states of Karnataka, Goa, Kerala and Maharashtra. *P. stocksii* is a medium sized bamboo which has stout, solid and strong culm. Culm attains a height of 9 m; internodes are 15-29 cm in length and 2.5 - 5.8 cm in dia. This species has great economic and ecological importance. It is used for construction, making baskets, umbrella handles and poles. Gregarious flowering was observed twice during 1884 and 1889 in North Kanara.

P. stocksii is mostly confined to the banks of streams and requires deep and well drained soil. It is cultivated in coastal belt of Karnataka and is considered as an important agroforestry species. *P. stocksii* is traditionally propagated using offset cutting. Culm cuttings require eight months to produce clonal plant with 90% success rate. Sporadic flowering in *P. stocksii* is common but seed setting has not been reported. In 1994, flowering has been observed in silent valley, Kerala.

Peculiar flowering habit in bamboo makes breeding for superior traits difficult. Selection and subsequent vegetative propagation of superior clones allow rapid improvement of bamboo germplasm while maintenance of broad genetic base improves scope of further improvement (Geilis and Oprins, 1998). Different propagation techniques available for bamboo include seed propagation, clump division, rhizome and culm cuttings (Banik, 1994).For mass scale propagation (>50000 plants/yr), classical techniques are inefficient (up to 10000 plants/yr) and tissue culture is the only reliable method (Geilis *et al.*, 2002).

Plant tissue culture work on bamboo species has been reviewed by INBAR (1991), Saxena and Dhawan (1994), Zamora (1994) and Woods *et al.* (1995).

Most of the studies on tissue culture regeneration of bamboo are based on either somatic embryogenesis through seedling explants (Rao *et al.*, 1990; Chambers *et al.*, 1991; Sood *et al.*, 1992; Chang, 1995; Woods *et al.*, 1995) or axillary shoot proliferation (Saxena, 1990; Arya and Sharma, 1998; Arya *et al.*, 1999; Ravikumar *et al.*, 1998). *In vitro* regeneration from explant of field grown mature culm have been reported with varying success rates in *B. bambos* (Arya and Sharma, 1998), *B. edulis* (Lin and Change, 1998), *D. giganteus* (Ramanayaka and Yakandawala, 1997), *D. strictus* (Chaturvedi *et al.*, 1993; Ravikumar *et al.*, 1998), *Phyllostachys viridis* (Hassan and Debergh, 1987) and *Thamnocalamus spathiflorus* (Bag *et al.*, 2000).

With the above background, a study was taken up for development of a protocol for rapid and mass clonal propagation of *P. stocksii* through axillary proliferation using explants from (Candidate Plus Clump).

MATERIAL AND METHODS

Surface sterilization of explants

Explants (nodal shoot segments) were surface sterilized using 70% ethanol (v/v) for 30-45 s, followed by 3-4 times wash with sterile distilled water and subsequent sterilization using 0.075 – 0.1% (w/v) mercuric chloride for 4-5 min depending upon explant type and size, followed by thorough (6-7 times) washing with sterile distilled water. Explants were then inoculated in various media for shoot initiation. Cultures for shoot initiation and multiplication were maintained at 28 ± 1°C under 2500 lux light with 12 h photoperiod.

Effect of liquid v/s semi-solid media on shoot induction

In order to achieve high rate of multiple shoot induction and subsequent growth, liquid and semi-solid media (Murashige and Skoog, 1962), supplemented with additives viz., ascorbic acid (50 mg/l), citric acid (25 mg/l), cysteine (25 mg/l), (0.1 mg/l) NAA and BAP (1.0 - 2.5 mg/l), were used.

Effect of growth hormones on shoot initiation

Various concentrations of auxins (IAA, NAA; 0.0 – 0.5 mg/l) and cytokinins (Kn, BAP; 0.5 – 2.5 mg/l), were tried for multiple shoot induction and subsequent shoot growth. Each treatment consisted of 12 replicate cultures and each experiment was repeated twice.

Effect of nutrient media on shoot induction and subsequent growth

In order to understand optimum nutritional requirement for high rate of multiple shoot initiation and subsequent shoot growth, various nutrient media, viz., B5 (Gamborg *et al.*, 1968), Hellers WP (Wood Plant medium); MS, MS/2 and MS/4 were used. Various liquid media, fortified with additives and growth hormones NAA (0.25 mg/l) and BAP (1.0 mg/l), were also used.

Shoot multiplication

In order to increase number of shoots for further studies on multiplication and *in vitro* rooting, single shoot/shoot clumps were excised from the mother explants and sub-cultured on fresh MS liquid medium with additives [NAA (2.5 mg/l) and BAP (1.0 mg/l)]. Sub-culturing on fresh medium was carried out at the interval of 2-3 weeks.

Effect of growth hormones on shoot multiplication

Two cytokinins (Kn and BAP) were tested in various concentrations with NAA (0.25 mg/l) in MS liquid medium with additives to standardize cytokinin concentration for further multiplication of *in vitro* differentiated shoots. In each culture bottle, 4 shoot clumps, each with 3-4 shoots, were used. Each treatment consisted of five replicates and each experiment was repeated twice.

A separate experiment was carried out in order to find out effect of various concentrations of NAA (0.25 mg/l), Kn and BAP in MS liquid medium with additives. Cultures were incubated at 28 $\pm$ 1°C temperature under 2500 lux light at 12 h photoperiod.

Effect of sucrose on multiplication and growth of shoots

To standardize optimum requirement of exogenous energy source for rapid shoot multiplication and subsequent growth, various concentrations of sucrose [0.0 – 6.0% (w/v)] were added in MS liquid medium fortified with additives [NAA (0.25 mg/l) and BAP (1.0 mg/l)]. 3 - 4 shoots/clump were used for the above experiment.

There were five replicates for each treatment and each experiment was repeated twice.

Inoculum size and density

In order to optimize size and density of inoculums, *in vitro* differentiated shoots were used as follows: single shoot (2.5 - 3.5 cm in length), 2-3 shoots clump, 3-4 shoots/clump, 5-6 shoots/clump and 7-8 shoots/clump. There were five replicates for each treatment.

In vitro rooting

In vitro multiplied shoots obtained after five subcultures were used for *in vitro* rooting. In order to ensure high rate of rooting and subsequent growth, single shoot and 2-3 shoot clump were isolated from the shoot multiplication cultures and used for the *in vitro* rooting. To find out the best auxin source and its concentration, various auxins, viz., IAA, IBA, NAA, NOA and 2, 4-D were tested with varying concentrations, viz., 0.1 - 1.0 mg/l (except 2, 4-D where it was 0.1-0.5 mg/l) in the MS/2 semi-solid medium. MS/4 medium was also tested using various auxins IAA, IBA, NAA and NOA (1.0 mg/l). In order to optimize nutrient medium for *in vitro* rooting, various media, viz., B5, WP, HE, MS, MS 3/4, MS ½ and MS/4, were tested. Media were also fortified with NAA (1.0 mg/l). For all the experimental media, pH was adjusted to 5.8 before autoclaving. Shoot clump with 2-3 shoots, were used as inoculum. There were five replicates for each treatment. Observations on percentage rooting, root number, root length and shoot length were recorded.

Ex vitro rooting

Two experiments, viz., effect of various auxins and their concentrations and rooting media, were conducted for *ex vitro* rooting. Single shoot as well as shoot clumps (2-3 shoots/clump) were used as inoculum.

In order to find out the best auxin and concentration for pulse treatment, various auxins, viz., IAA, IBA, NAA and NOA, were tested. Shoot clump (2-3 shoots/clump) treated with various auxin solutions were planted in soilrite treated with 0.1% (w/v) Bavistin in iron trays (118"x12"x6"). The trays, covered by plastic sheet to maintain high humidity, were kept inside mist chamber at 30 $\pm$ 5°C and 80 $\pm$ 5% RH. Plastic sheet was removed after 3-4 weeks and observations taken on various root parameters.

To find out the best rooting medium for *ex vitro* rooting various rooting media, viz., sand, soil and soilrite, either alone or in combination, were tested. Shoot clump of 2-3 shoot were used for *ex vitro* rooting. Shoot clumps were pulse treated with IBA (1000 ppm) for 5 min. In another experiment pulse treated shoot (4.0-5.0 cm), and shoot clump (2-3 shoots/clump) treated with IBA (1000 ppm for 5 min), were planted in sand bed raised over gravel. The entire set-up was covered with polythene sheet to maintain high RH. The experiment was conducted in mist chamber at 30 $\pm$ 5°C and 80 $\pm$ 5% RH.

Hardening of plantlets

In vitro rooted shoots were hardened by two methods. In the first approach, *in vitro* plantlets were washed with tap water and transferred to culture bottles with autoclaved and moistened hardening media (sand, soilrite and vermiculite, separately or in combination). Bottles were covered with caps and maintain under light at 28 $\pm$ 2°C for 3-4 weeks. In the second approach, *in vitro* rooted plantlets were directly transplanted into potting medium of soil, sand and compost (4:1:5 v/v) enriched with neem cake [10 kg/cu m + single super phosphate (SSP 2.4 kg/cu m)], fungicide (M-45, 0.4 kg/cu m) and Phorate (0.4 kg/cu m). Polybags (600 cc) and root trainers (600 cc) were used as containers. Initially, plantlets were kept under polyglobule for 2 weeks inside mist chamber. *Ex vitro* rooted shoots were directly transferred into potting medium in 600 cc poly bags and kept in mist chamber for 2-3 weeks and later on kept under shade for 2-3 weeks, before transfer open nursery.

Macroproliferation

About five months old plantlets, with 6-8 tillers and rhizomes, were used for macroproliferation. Shoots with rhizome and roots were transplanted into fresh potting medium in the polybags (600 and 1000 cc) for further growth. Initially, macroproliferated plants were kept inside mist chamber/shade for 3-4 weeks, and later under nursery condition, for further growth and development.

In vitro flowering

Cultures of shoot multiplication and somatic embryogenesis showing sporadic flowering were maintained separately. Different media and auxins were tested for *in vitro* flowering and cultures were maintained at 28 ± 1°C.

Field trial

Field trial of *in vitro* cloned plants of *P. stocksii* was conducted in Sept 2004 at Gottipura (Karnataka, India). Survival and growth performance were recorded up to six months. Further field trials were conducted during 2005-06.

RESULTS AND DISCUSSION

Effect of liquid or semi-solid medium

MS liquid medium proved better for high rate of multiple shoot induction with better shoot growth. In this medium, 4-6 shoots/explant was achieved within 3 weeks. Comparatively dwarf and lesser number of shoots (2-3 shoots/explants) were observed in MS semi-solid medium; leaching and browning were also observed in this medium.

Effect of size of explant on shoot induction

Nodal shoot segment of 2.5 - 3.0 cm in length and 0.25 - 0.35 mm in diameter proved to be the best for high rate of multiple shoot induction without problems of leaching or browning. Stout shoots (> 0.35 mm thickness) posed problems of leaching and browning; contamination also increased with size of explants. Culm segment induced more vigorous shoots than the culm branch nodal segment. However, no differences were observed in further multiplication rate and growth.

Effect of growth hormones on shoot initiation

MS liquid medium supplemented with additives, NAA (0.25 mg/l) and BAP (1.0 - 2.5 mg/l) proved better for multiple shoot initiation and shoot growth. Within 2-3 weeks, about 4-6 shoots were induced on this medium. Higher concentrations of BAP (2.5 mg/l) inhibited shoot growth. Kn, in all concentrations, induced lesser number of shoots. MS liquid medium supplemented with NAA (0.25 - 0.5 mg/l) and BAP (1.0 mg/l) induced 100 percent multiple shoots (6 shoots/explant) with shoot length of 3.66 - 4.17 cm (Fig. 1) from the nodal shoot segment within 3 weeks. Lower concentrations of BAP (0.5 mg/l) with NAA induced less number of shoots.

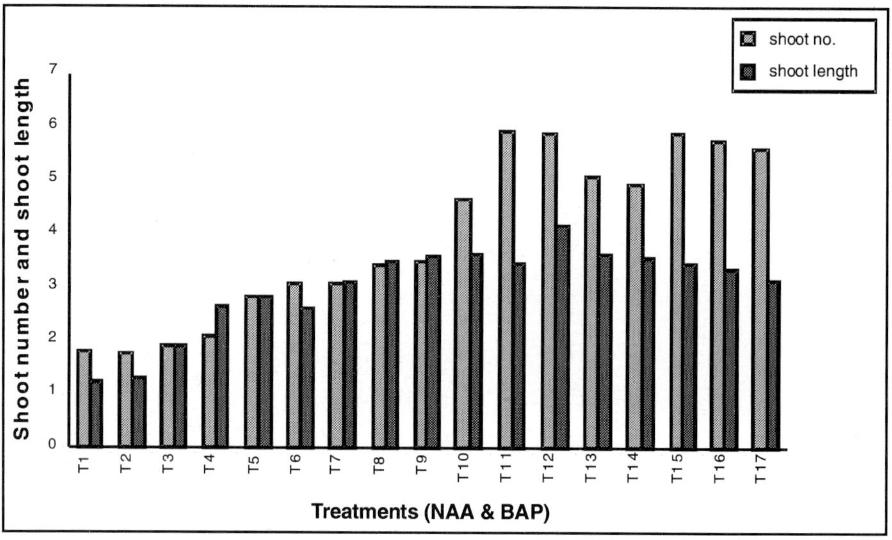

Fig. 1. Effect of plant growth regulators (NAA & BAP) on shoot initiation.

Effect of nutrient media on shoot induction

MS medium fortified with additives, NAA (0.25 mg/l) and BAP (1.0 mg/l) induced multiple shoots (4-5 shoots/explants) within 2 weeks followed by the B5 and WP media. Poor multiple shoot induction and growth was observed in Heller's medium. Liquid media proved better than semi-solid media in terms of number of shoots and shoot length. Further modification of MS medium did not improve number or length of shoots.

Shoot multiplication

Shoot clump with 2-3 shoots proved better for clump formation and multiplication of *in vitro* shoots on fresh shoot multiplication medium. On MS liquid medium with additives, NAA (0.25 mg/l) and BAP (1.0 mg/l), initial multiplication rate was 2-3 fold in 2 weeks.

Effect of growth hormones on shoot multiplication

Maximum shoots (36 clumps) were induced on MS medium with additives NAA (0.25 mg/l) and BAP (1.0 mg/l). High rate of shoot multiplication and growth was also observed in MS medium with additives, NAA (0.25 mg/l) and BAP (2.5 mg/l), but shoots were dwarf (Fig. 2). Incorporation of auxin (NAA) proved essential for shoot multiplication and particularly for shoot growth. Kinetin proved less effective as compared to BAP for shoot multiplication.

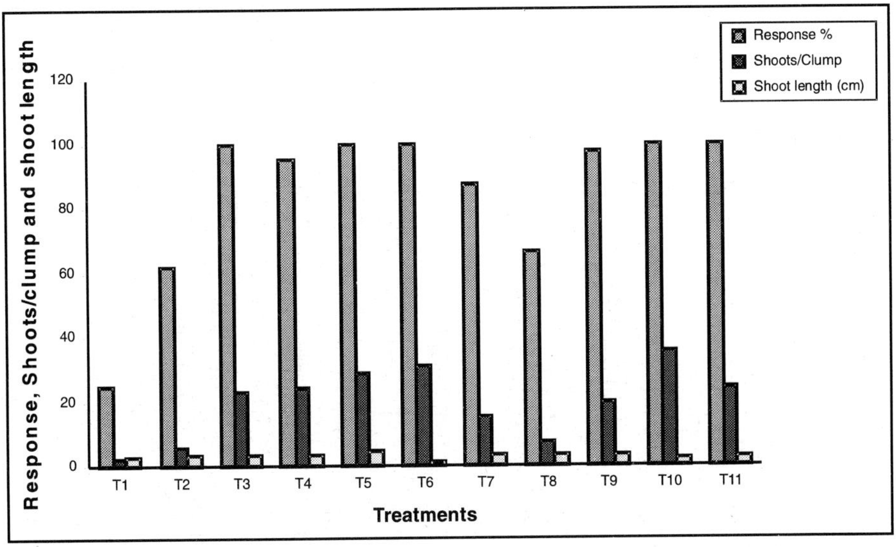

Fig. 2. Effect of plant growth regulators on shoot multiplication.

Effect of various nutrient media on shoot multiplication

MS liquid medium with additives, NAA (0.25 mg/l) and BAP (1.0 mg/l) proved the best, followed by WP and B5. Similar to shoot initiation results, MS/2, MS/4 and HE proved ineffective for shoot multiplication. MS liquid medium proved better (21 shoots/shoot clump) as compared to semi-solid medium (13 shoots/clump). Differences in shoot multiplication were also observed in liquid and semi-solid media (Fig. 3).

Effect of sucrose concentrations on the shoot multiplication

MS medium supplemented with 3.0% (w/v) sucrose proved the best for maximum shoot production (18.75 shorts/shoots/ clump) with average shoot length of 4.05 cm (Fig. 4). Medium without sucrose did not support survival of shoots, which revealed that *P. stocksii*, under *in vitro* condition, lacks even partial autotropism. Higher concentrations of sucrose did not improve shoot multiplication as well as growth.

Effect of inoculum size and density on shoot multiplication

4-5 shoots/clump proved better inoculum size for multiplication and subsequent growth in semi-solid medium (Fig. 5). Increase in number of shoots/clump did not improve multiplication rate and growth, which may be attributed to the nutrient competition for the

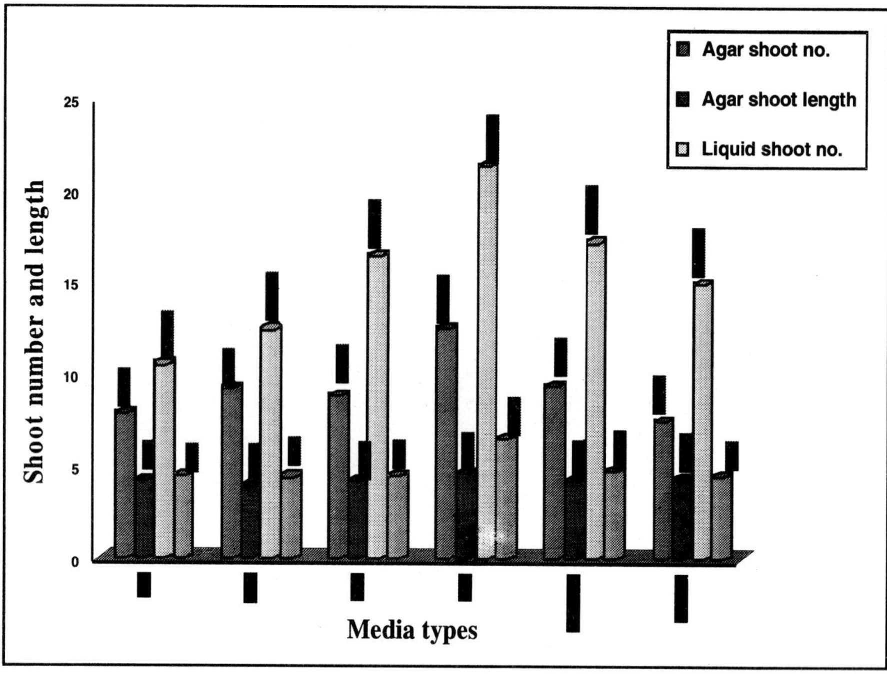

Fig. 3. Effect of various media on shoot multiplication.

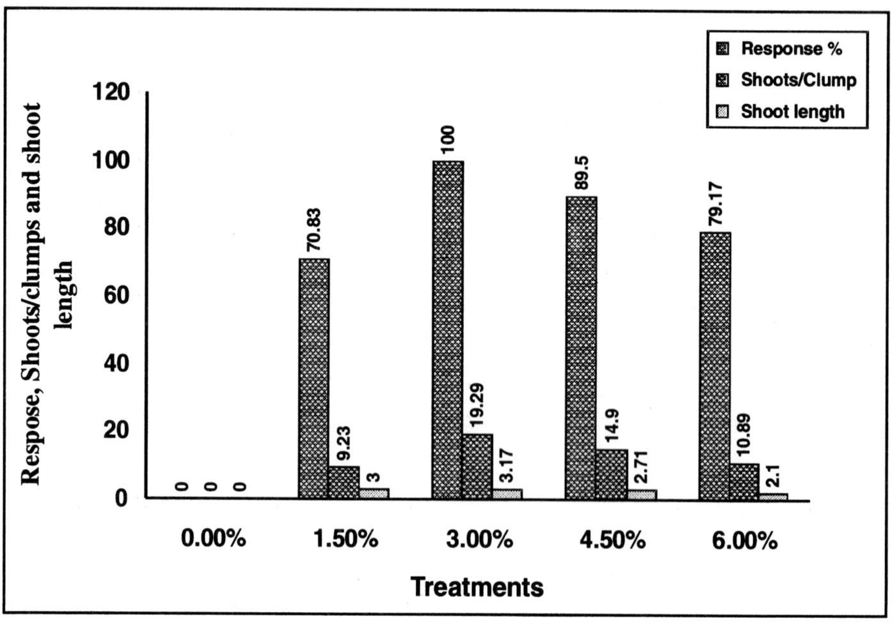

Fig. 4. Effect of sucrose concentration on shoot multiplication.

growth with increase in inoculum size. In semi-solid medium, 3-4 shoots/clump proved ideal inoculum density for shoot multiplication and growth.

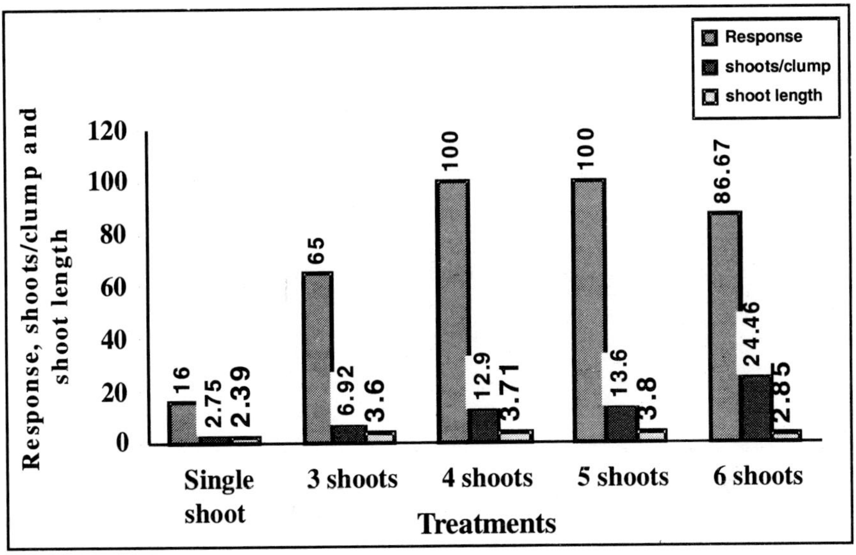

Fig. 5. Effect of inoculum size on shoot multiplication.

In vitro rooting

Among the single shoot and shoots clump (2-3 shoots clump), shoot clump was better for *in vitro* rooting. Shoot clump of 2.5 - 4.0 cm in length proved optimum for the rooting.

Effect of auxins on *in vitro* rooting

MS/2 medium and MS/4 medium with IAA, IBA, NAA or NOA (1.0 mg/l) and MS/4 medium with NAA (1.0 mg/l) induced 100% rooting. Significant differences were observed with respect to effect different auxins on rooting percentage, root number, root and shoot length. Lower concentrations of 2, 4-D (0.1-0.5 mg/l) favoured rooting and shoot growth (12.6 cm). However, in the above treatment, callus was observed from the shoot base, which is undesirable and pose problems at hardening phase. Among the auxins tested, IAA proved less effective (Fig. 6).

Effect of various nutrient media on *in vitro* rooting

MS/2 with NAA (1.0 mg/l) proved best for high frequency rooting and root number per clump. MS2 and MS/4 favoured more number of roots. In terms of induction of rooting, WP proved better than B5 and Heller's medium unsuitable for *in vitro* rooting (Fig. 7).

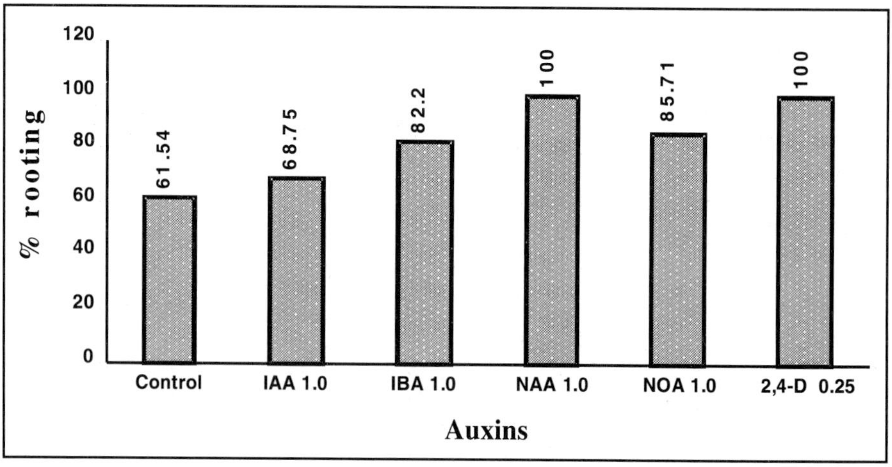

Fig. 6. Effect of various auxins on *in vitro* rooting.

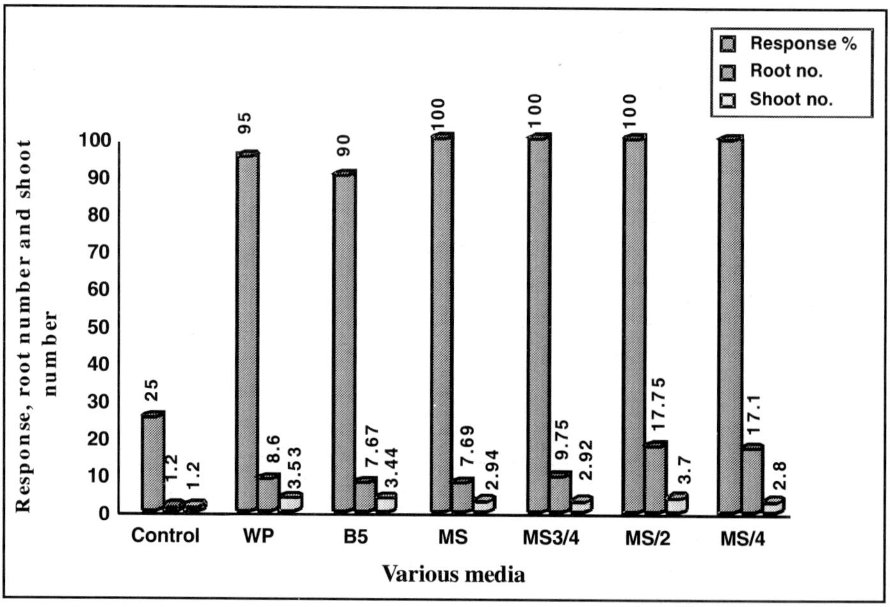

Fig. 7. Effect of various media on *in vitro* rooting.

Effect of auxins on *ex vitro* rooting

Shoot clumps, pulse treated with NAA (1000 ppm) followed by IBA, responded well in terms of rooting, root number per clump, root length as well as tiller length. Effect of auxin treatment was significant and increased rooting frequency. Minimum rooting frequency was observed with NOA followed by IAA (Fig. 8).

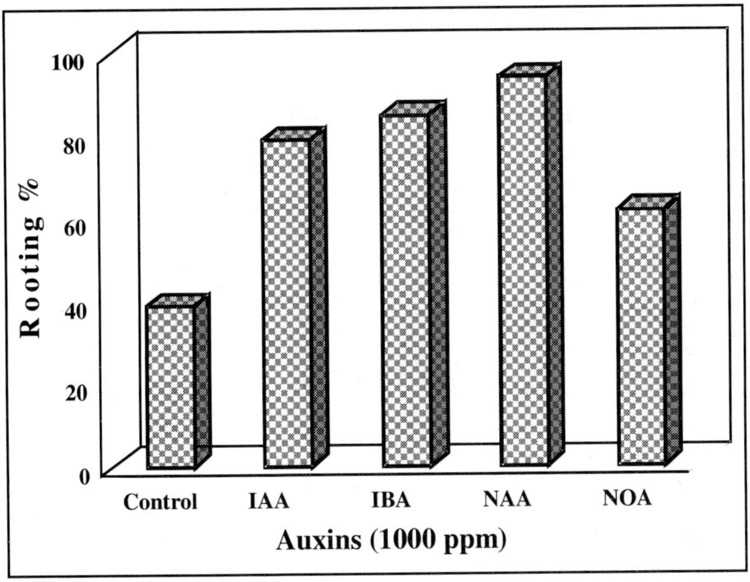

Fig. 8. Effect of various auxins on *ex vitro* rooting.

Effect of rooting media on *ex vitro* rooting

Shoots/clumps treated with NAA (1000 ppm) and planted in sand showed more rooting (99%) with 3-4 roots from each shoot clump, within 3-4 weeks. Combined use of soil, sand, soilrite and vermiculite, however, did not improve above growth parameters (Fig. 9).

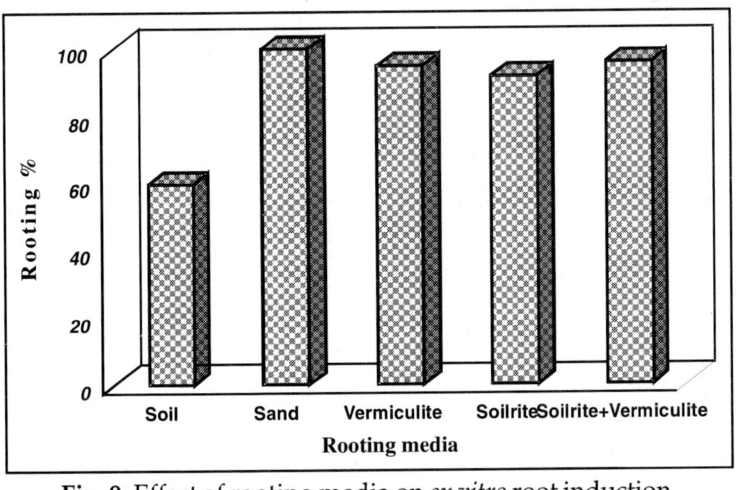

Fig. 9. Effect of rooting media on *ex vitro* root induction.

In vitro rooting and hardening

Among two approaches tried, the first approach viz., use of soilrite, sand and vermiculite as a hardening medium, increased survival rate. New root and shoot developed within two weeks. Such hardened plants exhibited 100 percent survival after transplantation into containers. However, after transplantation into containers, further hardening for 3-4 weeks was considered essential. This approach increased one more step in the hardening process and required two additional weeks to complete the entire production schedule.

In the second approach, 4-week-old *in vitro* rooted plants exhibited 95 percent survival after transplantation into polybags/root trainers and hardening for 4 weeks in mist chamber at 30 $\pm$ 5°C and 80 $\pm$ 5% RH.

Ex vitro rooting and hardening

Ex vitro rooting excluded one step of *in vitro* rooting thereby reducing cost of production and improved quality of plantlets. Direct *ex vitro* rooting in sand bed resulted in high rate of survival (> 95 %); plantlets maintained further growth phase in containers. Retention of transplanted plantlets for about 3-4 weeks in mist chamber favoured new root and shoot growth and miniature rhizome development.

Macroproliferation

Macroproliferation provides scope for large-scale multiplication of micropropagated planting material. From 5-6 month old plantlets, which had attained height of 20-30 cm with 5-6 tillers and 2-3 miniature rhizomes, 1-2 tillers with root and rhizome, were separated and further multiplied. Average multiplication rate through this method was 4 fold. Hardening in shade/mist chamber for 4-5 weeks favoured early establishment and high survival of the macro proliferated plants. From 500 explants, 2000 plants could be produced after 3 months. Taller plants (25-30 cm height) with 4-5 tillers, large number of roots and well developed rhizome, were suitable for field planting. Macroproliferation reduced cost of production and increased number of planting material.

In vitro flowering

Sporadic *in vitro* flowering was observed during shoot multiplication, root initiation somatic embryo production in *P. stocksii*

(Plate 2c, d). It was observed that 10-20% rooting cultures treated with NOA and 2,4-D, produced flowers directly from the rooted shoots. *In vitro* flowering was more in somatic embryo derived plants, which were grown in media with 2, 4-D as a growth hormone. However, these observations need more detailed study.

Production of plantlets

Based on the above protocol, more than 5000 plants were produced during 2002-05, through axillary shoot proliferation. Rooting percentage was about 100 percent (*in vitro* as well as *ex vitro* processes). This may be because of high level of rejuvenation and revigomation due to repeated *in vitro* shoot multiplication. In sand medium, 4-week-old rooted plantlets attained 8-10 cm height with 4-6 roots. Hardening in mist chamber for 4-5 weeks proved essential to promote new root growth, establishment of plants, shoot growth and high rate of survival. Acclimatization for further 3-4 weeks under shade before transfer to open nursery favoured better growth and survival. After *in vitro/ex vitro* rooting, clonal plantlets (20-25cm in height with miniature rhizome and minimum 4-5 tillers) could be produced within 4 months in 600 cc polybags/root trainer.

Field trials

Initial field trial conducted at Gottipura in the month of Sept 2004 exhibited 100 percent survival despite the trial laid during the off monsoon. Plantlets maintained growth phase, even after ten months of planting.

CONCLUSION

In the present study, through a rapid and simplified method of *in vitro* cloning of *P. stocksii,* high frequency multiple shoots (4-6 shoots/ explants) could be obtained in liquid MS medium with additives. Semi-solid medium also induced multiple shoots but fewer in number and poor in growth. With regard to season of collection of explant, March-July proved ideal for high frequency multiple shoot initiation with minimum problem of bacterial contamination.

In *P. stocksii,* 5 fold multiplication rate could be achieved using MS liquid medium with additives within 4 weeks period. High concentration of BAP improved multiplication rate but shoots were dwarf and unsuitable for multiplication in liquid medium as well as for rooting. Sub-culturing within two weeks in fresh shoot multiplication medium was found essential. Growth and vigour was maintained in *P. stocksii* for more than 2 years, without any sign of deterioration of multiplication rate and abnormality in shoots.

Low nutrient (MS/4 basal) medium with NAA (1.0 mg/l) exhibited 100 percent rooting from the 2-3-shoots/clump. High frequency (100 percent) rooting was also encountered from the shoot base. In the present study high frequency (> 95 percent) *ex vitro* rooting could be obtained by pulse treatment of NAA (1000ppm) followed by planting in sand bed medium in mist chamber. Sand, being a cheap and easily available medium, could be the best option for *ex vitro* rooting. However, for *ex vitro* rooting under open conditions where outside humidity is low, vermiculite and soilrite could be very useful rooting media since both these rooting media have high moisture retention capacity.

ACKNOWLEDGEMENT

Director and Group Co-ordinator (Research), Institute of Wood Science & Technology, Bangalore, provided facilities and constant encouragement to carryout research work.

References

Arya, S. and Sharma, S. (1998) Micropropagation technology of *Bambusa bambos* through shoot proliferation. *Indian For.* 124: 725-731.

Arya, S., Sharma, S., Kaur, R. and Arya, I.D. (1999) Micropropagation of *Dendrocalamus asper* by shoot proliferation using seeds. *Plant Cell Rep.* 18: 879-882.

Bag, N., Chandra, S., Palani, L. M. S. and Nandi, S. K. (2000) Micropropagation of Dev-ringal (*Thamnocalamus spathiflorus* (Trin) Munro) – a temperate bamboo and comparison between *in vitro* propagated plants and seedlings. *Plant Sci.* 156: 125-135.

Banik, R. L. (1994) Review of conventional propagation research in bamboo and future strategy. In: *Constraints to production of bamboo and rattan*, Technical Report No. 5. INBAR, New Delhi, pp. 115-142.

Chambers, S. M., Heuch, J. H. R. and Pirrie, A. (1991) Micropropagation and *in vitro* flowering of the bamboo – *Dendrocalamus hamiltonii* Munro. *Plant Cell Tissue Organ Cult.* 27: 45-49.

Chang, W.C. (1995) Somatic embryogenesis of *Bambusa oldhamii, Bambusa beecheaena* and *Sinocalamus latiflro*. In: *Somatic embryogenesis in woody plants* (eds. Jain, S. M., Gupta, P. K. and Newton, R. J.) Kluwer Academic Publishers, pp. 53-66.

Chaturvedi, H. C., Sharma, M. and Sharma, A. K. (1993) *In vitro* regeneration of *Dendrocalamus strictus* Nees through nodal segment taken from field grown culm. *Plant Sci.* 91: 97-101.

Gamborg, O. L., Miller, R. A. and Ojima, K. (1968) Nutrient requirements of suspension cultures of soybean root cells. *Exp. Cell Res.* 50: 151-158.

Geilis, J. and Oprins, J. (1998) The strategic role of biotechnology in mass scale production of woody bamboos. In: Sustainable agriculture for food, energy and industry (eds El Bassam, N., Bhel, R. K. and Prochnow, B.) James and James, London, pp. 165-171.

Geilis, J., Peeters, H. and Deberg, P. C. (2002) Tissue Culture strategies for genetic improvement of bamboo. In: Proceeding of the twentieth Intl. Eucarpia symposium section ornamentals - strategies for new ornamentals (eds. Van Huylebroeck, J., Van Bockstaele, E. and Deberg, P). *Acta Hort.* 552: 195-203.

Hassan, A. A. EI and Debergh, P. (1987) Embryogenesis and plantlet development in the bamboo - *Phyllostachys viridis* (Yound) MeClure. *Plant Cell Tissue Organ Cult.* 15 : 73-77.

Huang, L. C., Huang, B. L. and Chen, W. L. (1989) Tissue culture investigation of bamboo - IV. Organogenesis leading to adventitious shoot and plantlets on excised shoot apices. *Enviorn. Exp. Bio.* 29: 307-315.

INBAR (1991) Propagations of Bamboo and Rattan through tissue culture. INBAR, Delhi.

Lin, C. S. and Chang, W. C. (1998) Micropropagation of *Bambusa edulis* through nodal explants of field grown culms and flowering of regenerated plantlets. *Plant Cell Rep.* 17: 617-620.

Murashige, T. and Skoog, F. (1962) A revised medium for rapid growth and bioassay with tobacco tissue culture. *Physiol. Plant.* 15: 473-497.

Ramanayake, S. M. S. D. and Yakandawala, K. (1997) Micropropagation of the giant bamboo (*Dendrocalamus giganteus* Munro) from nodal explants of field grown culms. *Plant Sci.* 129: 213-223.

Rao, I. U., Ramanuja Rao, I. V., Narang, V., Jerath, R. and Pillai, K. G. (1990) Mass Propagation of Bamboo from Somatic Embryos and their Successful Transfer to Forest. In: *Bamboo Current Research - Proc. Intl. Bamboo Workshop*, KFRI, Peechi, pp. 167-172.

Ravikumar, R., Ananthakrishnan, G., Kathiravan, K. and Ganapathi, A. (1998) *In vitro* propagation of *Dendrocalamus strictus* Nees. *Plant Cell Tissue Organ Cult.* 52: 189-192.

Saxena, S. (1990) *In vitro* propagation of the Bamboo (*Bambusa tulda*.Roxb.) through proliferation. *Plant Cell Rep*. 9: 431-434.

Saxena, S. and Dhawan, V. (1994) *Micropropagation research in South Asia - Constraints to production of bamboo and rattan*. Technical Report 5, INBAR, Delhi.

Sood, A., Sharma, O. P. and Palni, L.M.S. (1992) Improved methods of propagation of maggar bamboo (*Dendrocalamus hamiltonii* Nees et Arn. Ex Munro) using single node taken from juvenile culm of elite seedlings. *J. American Bamboo Soc*. 9: 17-24.

Woods, S. H., Woods, J. E. and Collins, G. B. (1995) Somatic embryogenesis in bamboo. In: *Somatic embryogenesis in woody plants* (eds Jain, S.M., Gupta, P.K. and Newton, R.J.) Kluwer Academic Publishers, pp. 41-52.

Zamora, A. (1994) *Micropropagation research in South Asia - constraints to propagation of bamboo and rattan*. Technical Report 5.

❑❑❑

Forest Biotechnology in India © 2008
Editors: S.A. Ansari, C. Narayanan & A.K. Mandal, pp.183-188
Satish Serial Publishing House, Delhi

17

Development of *in vitro* Protocol for Plantlet Production in *Garcinia* A Tree Spice

Y.S RAO, K. MARY MATHEW, R. LAKSHMANAN, K.J. MADHUSOODANAN AND K. PRADIP KUMAR

Biotechnology Division, Spices Board (Ministry of Commerce & Industry, Govt. of India), P.B. No.2277, Palarivattam, Cochin-25, India

ABSTRACT

A micropropagation system for *Garcinia* was developed using axillary bud explants from mature trees. Murashige and Skoog's Medium (MS) at half strength supplemented with a combination of BA (5 mg^{-l}), GA_3 (0.5 mg^{-l}) and NAA (0.5mg^{-l}) was found suitable for the production of multiple shoots. Shoots were rooted on MS medium containing 1 mg^{-l} concentrations of IAA, IBA and NAA followed by culture on hormone free medium with activated charcoal. The present paper describes the protocol for *in vitro* plantlet production in *G. cambogia* and *G .indica.*

Key words: Camboge, *G. cambogia, G.indica,* Kokum, micropropagation.

INTRODUCTION

Garcinia spp. are distributed throughout the tropics and several of them have tremendous value for their spice and medicinal properties. Though there are more than 200 species in the genus, only 20 are found in India. Two species namely, *G. cambogia* (L.) Roxb. (common name: Camboge) and G. *indica* Choisy (common name: Kokum) are widely distributed in the Western Ghats and are very popular in south Indian cuisine. The fruit rind is reportedly the richest source of á-hydroxy citric acid, which also has metabolic regulatory function on obesity (Heymsfield *et al.,* 1998; Muthulakshmi and Sarah, 1999). Though the fruits have therapeutic values and extensively used in culinary preparations as a condiment, the tree genetic resources remains unexploited. *Garcinia* is generally propagated through seeds and grafts. The flowers are cross-pollinated and seeds are mono-embryonic. Dioecious nature of the trees, long pre-bearing period and prolonged seed dormancy impedes its propagation and improvement.

Unlike other crops, the problems faced during tissue culture of trees are plenty such as lack of reproducibility, low rate of rooting, phenolic exudation and high endogenous contamination. *Garcinia* spp. stand as a unique group and have always been a challenge to breeders and tissue culturists (Rao *et al.,* 1997). This paper deals with a micropropagation technique for rapid multiplication of Camboge and Kokum using explants from mature trees.

MATERIALS AND METHODS

For shoot multiplication, tender, actively growing shoots collected from approximately 20-year-old elite trees of *Garcinia* spps. (Camboge and Kokum) were used as explant source. Axillary buds/nodal segments from the shoots were collected, washed under running tap water for 1 hour and surface sterilised using 0.1% mercuric chloride. Subsequently, dissected nodal segments were suspended in sterile distilled water for 30 min to allow leaching out of harmful phenolic oxidation products. Murashige and Skoog medium (Murashige and Skoog, 1962) containing a variety of plant growth regulator combinations were employed for culture initiation and establishment. The plant growth regulators used were BAP, KN, 2iP, IAA, NAA and IBA. PVP was added to the medium for controlling phenolic exudations. Effect of adenine sulphate and phloroglucinol on shoot multiplication was also tested. The cultures were incubated at 16 hour photoperiod (2000 lux) at 28 ± 2°C.

RESULTS AND DISCUSSION

Experiments identified MS, with BAP (5 mg l^{-1}), GA (0.5 mg l^{-1}), NAA (0.5mg l^{-1}) and 5% sucrose, as the best medium suited for culture initiation and establishment for Camboge and Kokum (Table1; Fig.1).

On the above medium, average shoot multiplication ratio was 1: 3.5 for Camboge and 1:2.5 for Kokum. However, during sub –culture, slight exudation of polyphenols from the cut ends of shoot clusters into the media was observed. In order to minimize this, PVP was added at 0.1% concentration. Interestingly, PVP was also found to have positive effect on shoot development in Camboge (Table 1, Fig.1). Addition of phloroglucinol or adenine sulphate separately in the induction medium also gave similar results. Simultaneous shoot development and elongation could be observed on medium containing PVP (Fig. 1). Shoots of Camboge, approximately 3-4 cm length and with 2-3 leaves, were isolated and tested for *in vitro* rooting.

Table 1: Effect of different media on *in vitro* shoot multiplication in *Garcinia cambogia*.

S. No	Treatment	Mean no. of shoots	Mean length of shoots (cm)
1.	MS1 (½MS+BA+ NAA+GA)	3.5	1.13
2.	MS1+ Phloroglucinol	3.67	1.76
3.	MS1+ PVP	4.33	2.91
4.	MS1+ ADS	2.17	1.4

Fig. 1: Multiple shoots in kokum.

Maximum rooting efficiency was achieved by culturing shoots for 15 –20 days on ¼ strength MS liquid medium supplemented with 1 mg $^{-1}$ each of IBA, IAA and NAA and subsequently transferring them to auxin free medium with activated charcoal at 50 mgl^{-1} (Fig. 3). Number of roots varied from 1 to 3 per shoot and the rooting percentage was 65%. Rooted plantlets were primary hardened for 6 weeks in plastic cups filled with soil rite (Fig. 4) in humid chamber at 80% RH and 25^{0}C before transfer to polybags in secondary nursery (Fig. 2.4). Approximately 60% plantlets were successfully hardened.

Fig. 2: Multiple shoots in camboge.

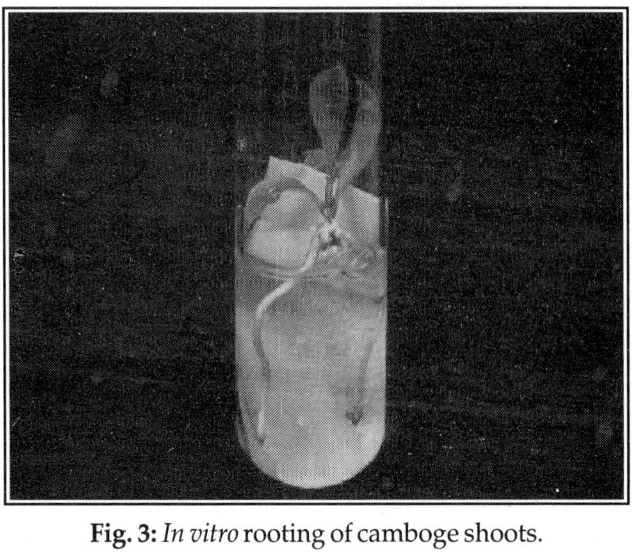

Fig. 3: *In vitro* rooting of camboge shoots.

Fig. 4: Primary hardened plantlet of camboge.

For *G. indica,* nodal explants of root sprouts from the 20-year-old elite trees were used. The media combinations used for the production of multiple shoots in *G. cambogia* were also found suitable for this species. However, multiplication rates in *G. indica* were slightly inferior when compared to that of *G. cambogia* and the ratio was 1:2.5. Multiple shoot proliferation could be observed within 15 days from culture initiation. Studies are being carried out for complete development of reproducible tissue culture protocol for both species as per strategy proposed in Fig. 5.

G. indica		*G. cambogia*
• Nodal segments from 20-year-old tress • 1/2 MS-BA (5) + NAA (0.5) +GA (0.5) • Time: 60-80 days • Sub-culture	**Multiple shoots & elongation**	• Nodal segments from 20-year-old tress • 1/2 MS-BA (5) + NAA (0.5) +GA (0.5) • Time: 60-80 days • Sub-culture
• 1/4th MS+IAA, IBA, NAA (each 1 mg) followed by HF medium + AC • Rooting%: 65 • No. of roots 1-3 • Length: 2-3 cm	**Rooting**	Trials are underway
• Soilrite in plastic cups • Humidity 80% • Temp: 25°C • Survival : 60%	**Hardening**	Trials are to be conducted

Fig. 5: Protocols for *in vitro* plantlet formation in *Garcinia* spp.

References

Heymsfield, S. B., Allison, D. B., Vasselli, J. R., Pietrobelli, A., Greenfield, D. and Nunez, C. (1998) *Garcinia cambogia* (hydroxy citric acid) as a potential antiobesity agent: a randomized controlled trail. *JAMA* 280 : 1569-1600.

Murashige, T. and Skoog, F. (1962) A. revised medium for rapid growth and bioassays with tobacco tissue cultures. *Physiol. Plant.* 15: 473-497.

Muthulakshmi, P. and Sarah, T.G. (1999) Kodampuli: a multipurpose fruit tree. *Indian Hort.* 44 (2): 7-8.

Rao, Y. S., Mary Mathew, K., Pradip Kumar, K. , Salykutty Joseph, Lakshmanan, R., Madhusoodanan, K.J. and Potty, S. N. (1997) Tissue culture studies on tree spices. In: *Proceedings of Biotechnology of Spices and Aromatic Crops*, pp 39-44.

□□□

Forest Biotechnology in India © 2008
Editors: S.A. Ansari, C. Narayanan & A.K. Mandal, pp.189-197
Satish Serial Publishing House, Delhi

18

Effect of Various Culture Conditions on *in vitro* Shoot Multiplication in *Bambusa nutans* Wall. and *Bambusa tulda* Roxb.

FATIMA SHIRIN, YOGESHWAR MISHRA, SUMAN YADAV, PRADEEP KUMAR PATEL AND S.A. ANSARI

Genetics and Plant Propagation Division, Tropical Forest Research Institute, P.O.-R.F.R.C., Mandla road, Jabalpur-482021, India

ABSTRACT

The shoot cultures of *Bambusa nutans* Wall. and *Bambusa tulda* Roxb. established through axillary buds of 10-year-old field grown clumps responded to various culture conditions such as basal media, culture phase and light intensity. In both species, basal media did not significantly influence shoot multiplication, but the interaction between WPM and 24 mM BA resulted in significantly high shoot multiplication in *B. tulda*. Better shoot multiplication was obtained on static cultures than cultures on rotary shaker in *B. nutans*. Two fold shoot multiplication was achieved in cultures exposed to 2.7 klux, which was significantly different from other light regimes in *B.tulda*.

Keywords : *Bambusa nutans, Bambusa tulda,* basal media, *in vitro,* shoot multiplication.

INTRODUCTION

World-wide interest in bamboo as a source of biomass in sustainable forestry systems has increased rapidly in recent years. In-spite of several classical propagation techniques, there exists shortage of planting material and depletion of natural resources. A lot of research has been focused on the development of micropropagation of bamboos throughout the year with axillary branching to meet colossal demand of true-to-type clonal planting stock. The present communication describes the *in vitro* plantlet propagation system for two commercially important bamboo species viz. *Bambus nutans* and *Bambusa tulda*. Both are economically important multipurpose bamboo species which flower gregariously after a long intermast period of 30-60 years. *Bambus nutans* is a medium sized bamboo used in Indian paper industry, for making poles, etc. *Bambusa tulda* is an evergreen gregarious bamboo used for covering houses, scaffolding and making pickles as well as for production of paper pulp and reinforcement in cement concrete (Tewari, 1992).

Various culture conditions viz., light and culture phase (static and shake cultures) are important factors in micropropagation of several species (Arditti and Ernst, 1992; Lal and Ahuja, 2004). However, reports on the effect of artificial light intensities and culture phases on *in vitro* growth of bamboos, are rather scarce. Bonga and Aderkas (1992) have discussed the suitability of a medium in terms of ionic strength, total nitrogen, ammonium/nitrate ratio, calcium deficiency and chloride sensitivity of the tissues. Therefore, we aimed at determining the most suitable media, light regime and culture phase for enhanced *in vitro* shoot multiplication of *Bambus nutans* and *Bambusa tulda*.

MATERIALS AND METHODS

Establishment of shoot cultures

Aseptic cultures were established using the surface sterilization procedure described in earlier publications (Mishra *et al.*, 2007; Yadav *et al.* unpublished).

The shoots established on MS (Murashige and Skoog, 1962) semi-solid (0.8% agar) medium supplemented with 3% sucrose and combination of 10μM BA and 3.0μM IAA for *B. nutans* and 10μM BA and 0.1μM IAA for *B. tulda* and passed through several subculture cycles were used for the present investigation.

Effect of basal media on shoot multiplication

Three liquid basal nutrient media containing the inorganic and organic salts of MS (Murashige & Skoog, 1962), B_5 (Gamborg *et al.*, 1968) and Woody Plant Medium (Llyod and McCown, 1980) were tried over four consecutive subculture cycles each of 15 days in *B.nutans.* BA (31.06 mM) and IAA (2.85 mM) was added uniformly to all three media. In *B.tulda* four basal media viz., MS, MS (NO^-_3), WPM and B_5 were evaluated to test their efficacy for shoot multiplication in combination with 0, 6, 12, 18 and 24 mM of BA.

Effect of phase of culture on shoot multiplication

In *B.nutans* two-way factorial experiment was carried out to compare the rate of shoot multiplication in static liquid cultures placed on culture racks and liquid cultures on the rotary shaker (100 rpm). The MS liquid medium was supplemented with three concentrations of BA (5, 10 and 20 mgl^{-1}). IAA (0.5 mgl^{-1}) was added uniformly to all treatments.

Effect of light regimes on shoot multiplication

In *B. tulda,* a propagule (bunch of three axillary shoots) was excised and inoculated on MS liquid medium supplemented with 100µM glutamine and 0.1µM IAA. Four different light intensity viz., 1.7 klux, 2.7 klux, 3.2 klux and 4.4 klux were evaluated for *in vitro* shoot multiplication. The rate of shoot multiplication and shoot length was determined after 15 days of shoot inoculation.

Culture condition and statistical analysis

The culture conditions have been described elsewhere (Mishra *et al.,* 2007; Yadav *et al.* unpublished). The data recorded for *in vitro* shoot multiplication rate was subjected to two-way (factor) analysis of variance and 'F' test for ascertaining level of significance. If the data were found significant at $P \leq 0.05$, $LSD_{0.05}$ was employed for comparison of treatment means.

RESULTS AND DISCUSSION

Effect of basal media on shoot multiplication

The subculture cycle in *Bambusa nutans* had a significant effect on the rate of shoot multiplication. In contrast, the basal media and their interaction with subculture cycle were found to be non-significant for *in vitro* shoot multiplication. The longest shoots (4.14 cm) were obtained on MS medium. The interaction between subculture cycle and basal media had no significant effect on shoot length (Table 1). Three media have different relative concentrations of NH_4^+ and NO_3^-. On par shoot multiplication in all three media may be because *B. nutans* can grow equally well on both high concentration medium like MS (20 meq l^{-1} NH_4^+) and low concentration medium like B_5 (2 meq l^{-1} NH_4^+). Results of this study are not in agreement with earlier workers who have found MS medium to be invariably more suitable than other media for shoot multiplication in bamboos (Saxena, 1990; Shirin *et al.*, 2003). In *Bambusa tulda* the rate of shoot multiplication remained on par in three basal media but was superior in all BA treatments over the control (Table 2). The interaction between BA concentrations and basal media had significant effect on shoot growth with WPM medium supplemented with 24µM BA resulting in maximum multiplica-

Table 1: Effect of subculture cycle, basal media and their interactions on shoot multiplication rate and shoot length in *Bambusa nutans*. Observations recorded after two weeks of inoculation.

	Subculture cycle (S)				
Basal Media (M)	1	2	3	4	Mean
		(a) Shoot multiplication (fold)			
MS	1.20	1.89	1.43	1.12	1.41
B5	1.10	1.47	1.28	1.08	1.27
WPM	1.30	1.55	1.46	0.79	1.23
Mean	1.20	1.64	1.39	1.00	
		(b) Shoot length (cm)			
MS	4.39	4.16	4.64	3.36	4.14
B5	3.32	3.38	3.26	1.94	2.97
WPM	3.17	3.94	4.20	2.62	3.48
Mean	3.62	3.83	4.03	2.64	

CD (0.05)	Shoot multiplication	Shoot length
M	NS	0.62
S	0.25	0.72
M * S	NS	NS

tion rate. Some woody species proliferate more rapidly on media containing WPM salts, resulting in higher number of shoots (Garton *et al.*, 1981; Goh *et al.*, 1988). Presumably, *in vitro* shoot growth in *B. tulda* needs low mineral nutrients. WPM medium is characterized by a lower potassium (K) and total nitrogen (N). Besides, shoot cultures of *B. tulda* may be more efficient in uptake and utilization of N and K so that lower levels of these elements in the medium are sufficient for shoot growth and development. Similar findings have been reported with some other woody species including *Alnus* (Garton *et al.*, 1981), *Salix* (Read *et al.*, 1982).

Table 2: Effect of basal media on shoot multiplication rate of *B.tulda*. Observations recorded after two weeks of inoculation.

Basal media (M)	BA μM (B)					
	0	6	12	18	24	Mean
MS	0.99	1.16	1.25	1.12	1.20	1.13
MS NO_3^-	0.53	0.93	1.00	0.62	0.72	0.77
WPM	1.01	1.25	0.88	1.06	1.48	1.14
B_5	0.87	0.90	0.85	1.19	1.10	0.99
Mean	0.85	1.04	0.99	1.00	1.13	
	CD (0.05)					
M	NS					
B	0.24					
M* B	0.52					

Effect of phase of culture on shoot multiplication

Stationary cultures of *B. nutans* exhibited significantly maximum shoot multiplication and length. The interaction between phases and BA concentrations was found to be on par for shoot multiplication (Table 3). BA and its interaction with phases had no significant effect on shoot multiplication and shoot length. It appears that the vitrification of shoots in liquid cultures kept on rotary shaker probably inhibit growth and development of shoots. In contrast to our results, other workers have reported increase in shoot growth using liquid shake culture (Agarwal and Gebhardt, 1994; Lal and Ahuja, 2004).

Effect of light regimes on shoot multiplication

In B. *tulda,* a maximum of 2 fold shoot multiplication with 2.7 klux, was significantly better than with other light regimes, but various light regimes did not significantly influence shoot length (Table 4, Fig. 1). Generally, higher light intensity costs more in terms of energy input as well causes stunting of the stem and leaf of alpine plants. Consequently, the results suggest that light intensity of 2.7 klux was sufficient to culture *B.tulda* economically. Soontornchainaksaeng *et al.* (2001) have also recommended moderate intensity of light for optimum shoot regeneration in *Phaius tankervilliae* and *Vanda coerulea.*

Table 3: Effect of phase, BA and their interactions on shoot multiplication and shoot length of *B. nutans*. Observations recorded after two weeks of inoculation.

	BA (mg/l)			
Phase (P)	5	10	20	Mean
	a. Shoot multiplication (fold)			
Stationary	2.10	1.06	1.62	1.73
Rotary Shaker	0.93	1.46	0.94	0.98
Mean	1.58	1.27	1.20	
	b. Shoot length (cm)			
Phase	5	10	20	Mean
Stationary	4.11	4.44	3.79	4.11
Rotary Shaker	3.55	3.34	2.90	3.26
Mean	3.83	3.89	3.35	

Variables	CD (0.05) Shoot multiplication	Shoot length
P	0.36	0.58
B	NS	NS
P*B	NS	NS

Table 4: Effect of light regimes on shoot multiplication and shoot length of *B.tulda*. Observations recorded after two weeks of inoculation.

Light regimes (klux)	Shoot multiplication (fold)	Shoot length (cm)
1.7	1.85	4.67
2.7	2.02	4.64
3.2	1.38	4.53
4.4	1.42	4.17
CD (0.05)	0.47	NS

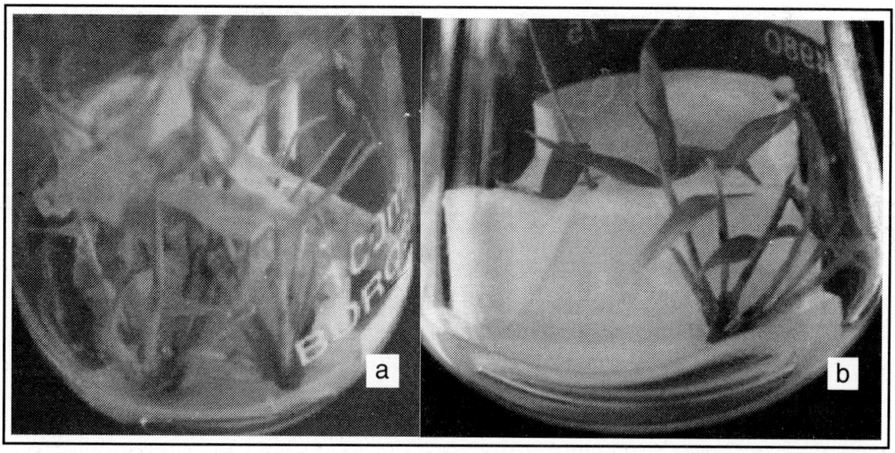

Fig. 1: Shoot multiplication: (a) *Bambusa nutans* (MS basal medium supplemented with 31.06 mM BA and 2.85 mM IAA) and (b) *Bambusa tulda* (WPM basal medium supplemented with 24 μM BA)

ACKNOWLEDGEMENT

The authors gratefully acknowledge the receipt of research grant from Department of Biotechnology, Government of India, New Delhi (Sanction No. 102/IFD/San 2632/2003-04) for the work reported herein.

References

Agrawal, D. C. and Gebhardt, K. (1994) Rapid micropropagation of hybrid willow (salix) established by ovary culture. *J. Plant Physiol.* 143 : 763-765.

Arditti, J. and Ernst, R. (1992) *Micropropagation of Orchids.* John Wiley & Sons Inc, USA.

Arya, I.D and Arya, S. (1997) *In vitro* culture and establishment of exotic bamboo *Dendrocalamus asper. Indian J. Exp. Biol.* 35 : 1252-1255.

Arya, I. D., Satsangi, R., Arya, S. (2001) Rapid micropropagation of edible bamboo *Dendrocalamus asper. J. Sus. For.* 14 : 103-114.

Bag, N., Chandra, S., Palni, L. M., Nandi, S. K. (2000) Micropropagation of Devringal. *Plant Sci.* 156 : 125-135.

Bonga, J. M. and Von Aderkas, P. (1992) *In vitro Culture of Trees.* Kluwer Academic Pub., Dordrecht.

Das, P. and Rout, G.R. (1991) Mass multiplication and flowering of bamboo *in vitro. Orissa J. Hort.* 19 (1-2) : 118-121.

Gamborg, O.L., Miller, R.A. and Ojima, K. (1986) Nutrient requirements of suspension cultures of soyabean root cells. *Exp. Cell.Res.* 50 : 151-158.

Garton, S., Hosser, M.A., Read, P.E. and Farnham, R.S. (1981) *In vitro* propagation of *Alnus glutinosa* Garten. *Hort Sci.* 16 : 758-759.

Goh, H.K.L., Rao, A.N. and Loh, C.S. (1988) *In vitro* plantlet formation in mangosteen (*Garcinia mangostana* L.). *Ann. Bot.* 62 : 87-93.

Gomez, K.A. and Gomez, A.A. (1984) *Statistical Procedure for Agricultural Research.* John Willey and Sons, New York, pp. 680.

Lal, N. and Ahuja, P. (2004) Assessment of liquid culture procedures for *in vitro* propagation of *Rheum emodi. Plant Cell Tissue Organ Cult.* 34:223-226.

Llyod, G. and McCown, B. (1980) Commercial feasible micropropagation of mountain laurel, *Kalmia latifolia*, by use of shoot tip culture. Com. *Pro. Intl. Plant Prop. Soc.* 30 : 421-427.

Mishra, Y., Rana, P. K., Shirin, F. and Ansari, S. A. (2001) Augmenting in vitro shoot multiplication by vipul and rhizogenesis by rice bran extract in *Dendrocalamus strictus. Indian J. Exp. Biol.* 39: 165-169.

Mishra, Y., Patel, P.K., Yadav, S., Shirin, F and Ansari, S. A. (2007) A micropropagation system for cloning of *Bambusa tulda* Roxb., *Sci. Horti.* (In press).

Murashige, T. and Skoog, F. (1962) A revised medium for rapid growth and bioassay with tobacco tissue culture. *Physiol. Plant.* 15: 473-497.

Ogita, S. (2005) Callus and cell suspension culture of bamboo plant *Phyllostachys nigra. Plant Biotech.* 22:119–125.

Ravikumar, R., Ananthakrishnan, G., Kathiravan, K. and Ganapathi, A. (1998) *In vitro* shoot propagation of *Dendrocalamus strictus* Nees. *Plant Cell Tissue Organ Cult.* 52:189-192.

Read, P. E., Garton, S. M., Louis, K. A. and Zimmerman, E. S. (1982) *In vitro* propagation of species for bioenergy plantation. In: *Plant Tissue Culture* (ed Fuziwara A.) Proc. 5th Intl. Cong. Plant Tissue and Cell Culture, The Japanese. Assn. Plant Tiss. Cult., Tokyo, Japan, pp 757-758.

Saxena, S. (1990) *In vitro* propagation of the bamboo *(Bambusa tulda* Roxb.) through shoot proliferation. *Plant Cell Rep.* 9:431–434.

Shirin, F., Arya, S. and Arya, I. D. (2003) Effect of nutrient media on *in vitro* shoot multiplication and rooting in *Bambusa vulgaris. Indian J. Trop. Biodiv.*, 11: 57-62.

Soontornchainaksaenga, P., Chaicharoenb, S., Sirijuntarutb, M. and Kruatrachue, M. (2001) *In vitro* Studies on the Effect of Light Intensity on Plant Growth of *Phaius tankervilliae* (Banks ex L' Herit.) Bl. and *Vanda coerulea* Griff. *Science Asia* 27: 233-237.

Tewari, D. N. (1992) *A Monograph on Bamboos.* International Book Distributors, Dehradun.

❑❑❑

Forest Biotechnology in India © 2008
Editors: S.A. Ansari, C. Narayanan & A.K. Mandal, pp.199-203
Satish Serial Publishing House, Delhi

19

Direct Shoot Regeneration from Embryonic Axis of *Oroxylum indicum* (L) Vent.

MAMTA GOKHALE AND Y.K. BANSAL

Plant Tissue Culture Lab., Dept. of Bioscience
R.D. University, Jabalpur, India

ABSTRACT

Embryonic axis explants of *Oroxylum indicum* were inoculated on MS medium supplemented with cytokinins *viz.*, BAP and KIN, alone and in combination. Maximum shoot length with root formation was induced on KIN (1.0 mgl^{-1}). Higher concentrations of BAP (5-10 mgl^{-1}) induced multiple shoot formation.BAP (1 mgl^{-1}) with KIN (0.1 mgl^{-1}) induced maximum number of shoots (5.50).

Keywords **:** BAP, cytokinin, KIN, *Oroxylum indicum.*

INTRODUCTION

Oroxylum indicum, commonly known as shivnak, is a small to medium sized tree which can reach up to 12 m height (Bennet *et al.,* 1992). Fruits of this tree are long woody capsules and each capsule contains numerous papery, flat, winged and light seeds. Every part of this tree possesses medicinal value. Dichloromethane extract of the stem-bark and root possesses antimicrobial, antifungal, anti-inflammatory and anticancerous properties (Ali *et al.*, 1998; Chen *et al.*, 2002). Root extract of this tree is an important the component in the best-known health care product 'Dashmula', an ayurvedic medicine. Besides, extracts of this species constitute one of the ingredients in chyawanprash (Vaidya, 1975). Adventitious shoot regeneration in this species has been recently reported (Gokhale and Bansal, in Press). The present paper reports regeneration of shoots directly from embryonic axisof *O. indicum.*

MATERIALS AND METHODS

Embryonic axis was used as explants in present experiment. Mature pods of the tree were collected from the forests of Chhindwara (M.P.), washed with running tap water, then treated with Teepol R (detergent) solution with continuous shaking for 5 min followed by repeated washing with sterilized distilled water. The seeds were then sterilized with 0.1% (w/v) $HgCl_2$ for 2-3 min and washed with sterilized distilled water (4-5 washings) to remove traces of $HgCl_2$. Embryonic axes, excised from the seeds, were inoculated on MS (Murashige and Skoog, 1962) medium (MSM) supplemented with different concentrations (0.1, 0.5, 1.0, 5.0 mg/l) of cytokinins viz., BAP, kinetin, alone or in combinations.

The pH of the medium was adjusted to 5.6-5.8, added with 0.80% (w/v) agar and autoclaved. The cultures were maintained at 25 ± 2^0C temperature with 14 h light photoperiod (2000 lux) and 60-65% RH.

RESULTS & DISCUSSION

Shoot induction

Effect of BA on multiple shoot induction

BA is the most widely used cytokinin for inducing multiple shoot formation through repeated divisions resulting into supernumerary vegetative buds. In the present experiment, high concentration (5 mgl^{-1}) of BPA induced multiple shoot formation directly from the explant (Table 1; Fig.1a-b) whereas low concentration (0.1 - 1 mg^{-1}) resulted in the formation of only a single shoot (Table 1). The shoot (s) induced were green and healthy and attained 6.5 cm height within 60 days of culture (Fig.1c).

Effect of kinetin

In the present experiment, kinetin was observed to be partially effective where its lower concentrations (0.1 – 1 mgl^{-1}) induced single shoots within 15 days of culture. However, higher concentrations (> 1 mgl^{-1}) proved ineffective (Table 1).

Table 1: Effect of BAP and kinetin on direct shoot formation in *Oroxylum indicum.*

	PGRS (mgl^{-1})	Frequency of shoot formation (%)	No. of shoots (Mean±SE)	Shoot length (cm)	Rooting*
Kn	0.1	70	1.00± 0.00	1.40± 0.30	+
	0.5	60	1.00± 0.00	3.50± 0.28	+
	1.0	60	1.00± 0.00	8.50± 0.28	+
	5.0	30	No response	No response	+
BAP	0.1	70	1.00± 0.00	0.76± 0.14	+
	0.5	70	1.00± 0.00	1.86± 0.20	-
	1.0	80	1.00± 0.00	2.80± 0.44	-
	5.0	40	4.60± 1.20	6.50± 0.44	-
BAP +Kn	0.1+0.1	70	1.00± 0.00	1.70± 0.13	+
	0.1+0.5	70	1.00± 0.06	2.30± 0.60	+
	0.1+1.0	70	1.00± 0.57	2.00± 0.50	+
	0.1+5.0	50	1.30± 0.33	0.90± 0.17	+
	0.5+0.1	60	1.00± 0.06	2.00± 0.28	-
	0.5+0.5	60	1.00± 0.88	1.50± 0.57	-
	0.5+1.0	60	1.00± 0.57	2.50± 0.28	-
	0.5+5.0	40	1.66± 0.66	1.50± 0.28	-
	1.0+0.1	60	5.50± 1.73	2.60± 0.92	-
	1.0+0.5	60	4.30± 2.73	2.60± 0.92	-
	1.0+1.0	70	4.00± 1.15	3.06± 1.01	-
	1.0+5.0	30	1.30± 0.33	0.50± 0.11	-
	5.0+0.1	40	4.80± 0.88	1.60± 0.54	-
	5.0+0.5	30	3.40± 0.66	2.00± 0.25	-
	5.0+1.0	30	2.30± 0.66	1.90± 0.5	-
	5.0+5.0	30	No response	No response	-

* +, Present; -, Absent

Effect of BAP and kinetin

Direct multiple shoot formation was observed after combined treatment involving both BAP and Kinetin (1.0 and 0.1 mgl^{-1}

respectively). Shoots regenerated were normal with bright green pigmentation and vigorous growth (Table 1).

Shoot Elongation

Rate of shoot elongation was faster in medium containing kinetin than in the medium with BAP. Maximum shoot length (8.5 cm) was observed on medium with 1.0 mgl^{-1} kinetin within 60 days of culture.

Rooting

Although BAP stimulates multiple shoot formation, it interferes with development and elongation of roots (Wright *et al.*, 1986). Individual shoots separated from multiple shoot clumps had to be subcultured on a plant growth regulator free MS medium (MS Basal medium). In some cases, lower concentrations of kinetin (0.1 – 1.0 mgl^{-1}) also induced rooting (Table 1; Fig.1d). The roots were thick, branched and creamish in colour. However, this trend was not observed with BA. All concentrations of BA proved unsuitable for root formation in *in vitro* regenerated shoots.

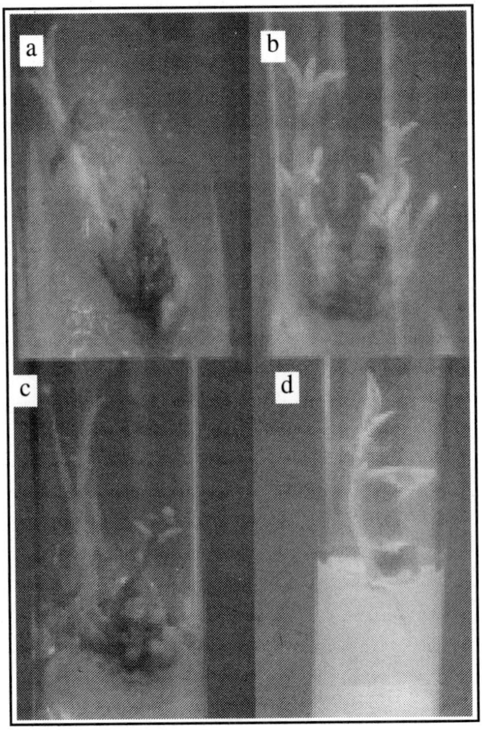

Fig. 1 a, b - Multiple shoot formation from embryonic axis; c - shoots after 50 days of culture; d - Root induction using kinetin

References

Ali, R. M., Houghton P. J. and Hoult J. R. S. (1998) Antimicrobial and anti-inflammatory activities of extracts and constituents of *Oroxylum indicum* (L.) Vent. *Phytomedicin*es 375-381.

Bennet, S. S. R., Gupta P. C. and Rao R. V. (1992) *Venerated plants.* ICFRE, Dehra Dun, pp. 147-149.

Chen, L. J., Games D. E and Jones J. (2002) Isolation and identification of four flavonoid constisuents from the seeds of *Oroxylum indicum* by HSCCC (High speed counter current chromatography). *J. Chromatogr. A* 988 : 95-105.

Gokhale, M. and Bansal, Y.K. Rapid *in vitro* regeneration of a medicinal tree species (*Oroxylum indicum* (L.) Vent. *Indian J. Trop. Biodiv.* (*In Press*)

Murashige, T. and Skoog F. (1962) A revised medium for rapid growth and bioassay with tobacco tissue culture. *Physiol. Plant.* 15 : 473-497.

Vaidya, B. G. (1975) Some controversial drugs of Indian medicine. *J. Res. Indian Medicine* 10 (4): 127.

Wright, M. S., Cairnes, M. G., Hinchee, M. A., David G. C., Kochler, S. M. and Williams, M. H., Colburn S. M. and Pierson, M. (1986) Plant regeneration from tissue culture of soybean by organogenesis. In: *Cell Culture & Somatic Cell Genetics of Plants*, Vol. 3 (eds Parrot, W. H. and White, E. L.) Academic Press, London, pp. 111-119.

❑❑❑

Forest Biotechnology in India © 2008
Editors: S.A. Ansari, C. Narayanan & A.K. Mandal, pp.205-213
Satish Serial Publishing House, Delhi

20

In Vitro Shoot Proliferation From Nodal Explants of Mature Trees of Kusum *(Schleichera oleosa)*

ANIMESH SINHA AND SHIREEN AKHTAR

Institute of Forest Productivity, Lalgutwa, Ranchi – 835303, Jharkhand

ABSTRACT

Kusum (*Schleichera oleosa*) is one of the most important lac-host forest tree species, which produces the best quality of lac having light colour as well as high yield per tree in India. Export of lac plays an important role in national trade. Hitherto, this tree has got very little attention for its improvement. In the present study, an attempt was made for direct regeneration of shoots from nodal explants of mature trees. Contamination and browning of culture were the major problems encountered during *in vitro* propagation. Contamination was high in rainy season in comparison to summer and winter seasons and was effectively controlled by treating the explants with 0.1% Bavistin and 0.1% mercuric chloride. Browning of cultures and media, due to secretion of excess phenols, was controlled by pretreating explants with chilled ascorbic acid solution (500 mg L^{-1}) and addition of ascorbic acid (10 mg L^{-1}), citric acid (10 mg L^{-1}) and PVP (50 mg L^{-1}) to the medium. Shoot initiation was achieved on Murashige and Skoog (MS) medium supplemented with cytokinin. Healthy shoot growth occurred on MS medium supplemented with BAP (5 mg L^{-1}), kinetin (5 mg L^{-1}), $AdSO_4$ (50 mg L^{-1}) and activated charcoal (2 g L^{-1}).

Keywords: Ascorbic acid, kusum, lac -host, phenol, sterilization, *Schleichera oleosa*.

INTRODUCTION

Schleichera oleosa (Lour.) Oken (Syn. *S. trijuga* Willd.) is a commercially important tree species belonging to the Sapindaceae family. In *Hindi* it is commonly known as *Kusum*. This tree helps produce the best quality lac with light colour and high yield. *Kusum* kernels yield a good quality oil used generally as hair oil, body massage oil and lubricant. It is also used in soap industries. Young fruits are pickled, and seeds and ripe fruits are eaten by tribals (Pradeepkumar and Prathapasenan, 2003). Peltit *et al.* (2000) isolated Schleicherastatins and Schleicheols from the bark and stem of *Kusum* tree, which are effective in curing cancer. Das and Chattopadhyay (2003) reported that the bark and seed oil of *Kusum* are effective in curing tuberculosis and arthritis, respectively.

Thus, the tree possesses a wide spectrum of nutritional and medicinal properties and deserves large-scale plantation.However, the tree has received little attention for its genetic improvement. Micro-propagation through nodal bud culture is one of the best alternative methods producing quality planting stock on a large scale within short time in each seasons. The present investigation was initiated with a view to developing protocol for *in vitro* propagation of *Kusum* using nodal explants of mature trees.

MATERIAL AND METHODS

The investigation was carried out at tissue culture laboratory of Institute of Forest Productivity, Jharkhand. Explants were collected from selected middle aged (30 - 40 years) trees of *Kusum* on the basis of criteria described by Sinha (2005). Explants of single nodal segments (2 - 3 cm length) were collected and thoroughly washed under running tap water for 1h to remove adhered dust and debris. The explants were dipped in 0.1% Bavistin solution for 20 min followed by rinsing with distilled water. Surface sterilization of explants in Bavistin with agitation using shaker and use of few drops of Tween 20 in sterilization process, were also tried. The explants were then surface sterilized using two sterilants individually for different duration of time. Finally, the explants were thoroughly washed with sterilized distilled water for 5 times (approx. 2 min each time) to remove traces of sterilants. Contamination rate in different seasons was also recorded. Like other woody species, the explants posed problem of browning. Nine different approaches were tried to control browning of explants (Table 1).

Table 1: Details of treatments followed for controlling of browning of *S.oleosa* cultures

Treatment	Remarks
BC_1	24 h dark followed by 16-h photoperiod + quick subculturing
BC_2	Culture in liquid medium
BC_3	Sealing the cut ends with paraffin wax
BC_4	Addition of activated charcoal in (0.2%) medium
BC_5	Pretreatment of explants with AA (0.1%) for 1 h
BC_6	Pretreatment of explants with AA (50 mg L^{-1}) for 1 h + Culture in liquid medium
BC_7	Pretreatment of explants with AA (50 mg L^{-1}) for 1 h + Addition of activated charcoal in (0.2%) medium
BC_8	Pretreatment of explants with AA (0.1%) for 1 h + addition of glutamine (400 mg L^{-1}) and casein hydrolysate (100 mg L^{-1}) in medium
BC_9	Pretreatment of explants with AA (500 mg L^{-1}) for 1 h + addition of AA (10 mg L^{-1}), CA (10 mg L^{-1}) & PVP (50 mg L^{-1}) in medium

Explants were trimmed at proximal and distal ends to remove dead tissues and 1.5 -2.0 cm long explants were inoculated into two different nutrient media, i.e. MS (Murashige and Skoog,1962) and WPM (Lloyd and Mc Cown, 1980) for tissue culture establishment. The media were prepared without plant growth regulators, but 3% sugar was added. For shoot initiation and proliferation, explants were subcultured on MS medium supplemented with different concentrations of cytokinin and auxin, alone or in combination. For root initiation, auxin treatments were tried. The pH of the medium was adjusted to 5.6 to 5.8 before adding agar (0.8% w/v) and autoclaving at 15 psi. The cultures were maintained at 25±2°C temperature with 16 h photoperiod. Survival, contamination of explants, percentage of bud sprouting and length of shoots were recorded.

RESULTS

The explants of *Kusum* were heavily contaminated within 3 days after inoculation. The effects of two sterilants, viz., $HgCl_2$ (0.1%, w/v) and NaOCl (3%, v/v) on controlling contamination are shown in Table

2. $HgCl_2$ showed better performance than NaOCl. Among durations of $HgCl_2$ application, 8 min application significantly reduced contamination and increased survival rate of explants (>60%). Contamination was the highest during rainy season and the lowest in summer (Fig. 1).

Table 2: Effect of two surface sterilants on producing aseptic culture of *S. oleasa* after 7 days of inoculation.

Sterilants	Duration (min)	Explants free from contamination (%)
$HgCl_2$ (0.1%)	2	0
	5	8.8
	8	63.2
	10	87.5
NaOCl (3%)	5	0
	10	12.2
	15	15.3

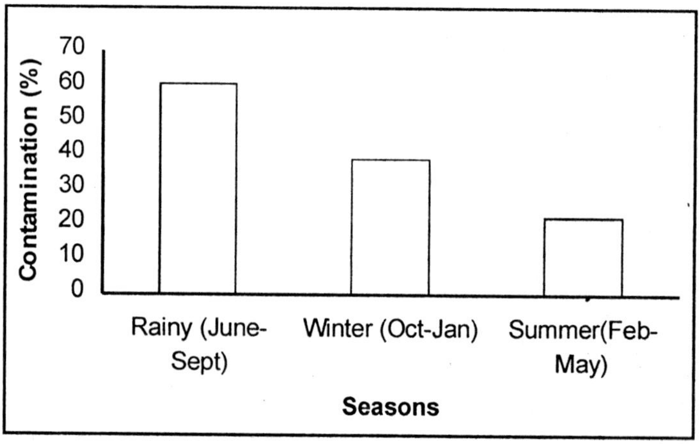

Fig. 1: Contamination of culture of *S. oleasa* during different seasons.

Nine different treatments (BC_1 to BC_9) were tried to control browning. The treatment of BC_1, BC_2, BC_3 and BC_4 showed very poor performance in controlling browning of *Kusum* cultures. Pretreatment of explants with antioxidant particularly AA, was essential for prevention of browning of cultures. Pretreatment of explants (treatment BC 9) before inoculation and also inclusion of antioxidants in inoculation medium, resulted in maximum number of cultures without browning (Fig. 2).

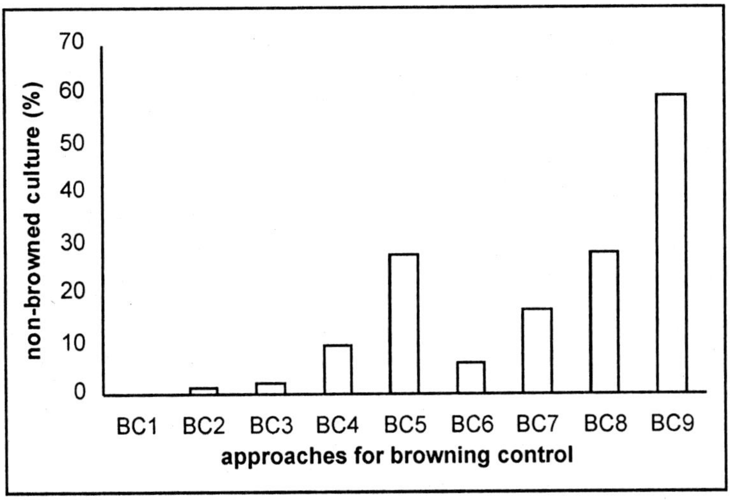

Fig. 2: Effect of different treatments on control of browning of *S. oleasa.*

Among the two basal media tested, MS medium proved to be the best (Fig. 3). As there was no growth regulator in basal media, bud sprouting or shoot initiation were not observed. After 4-6 weeks, explants were subcultured on MS medium containing cytokinin. Shoot initiation was observed in MS medium containing BAP (1, 3 and 5 mg L^{-1}). For better shoot proliferation, different growth regulators, viz., $AdSO_4$ and activated charcoal (AC) were added to the medium (Table 3). MS medium supplemented with BAP (5 mg L^{-1}), Kinetin (5 mg L^{-1}), $AdSO_4$ (50 mg L^{-1}) and AC (2 g L^{-1}) supported healthy growth of shoots. Later, elongated shoots were subcultured in MS media supplemented with auxins (IBA and IAA), alone and in combination. However, no rooting was observed (Table 3, Fig.3).

DISCUSSION

The primary objective of the present study was to generate large numbers of propagules with desirable characteristics. Sexual reproduction is not a viable strategy because of the low seed viability, poor germination rates or genetically non-uniform offspring. On the other hand, micropropagation of mature trees provides multiplication and evaluation of elite trees (McGranahan *et al.*, 1987).

Table 3: Effect of growth regulators on proliferation of shoots after 3 weeks of first subculture of *S. Oleosa*.

Treatment	Shoot length (cm)	No. of branches > 1 cm length	No. of leaves	No. of broad leaves
BAP (5 mg L^{-1})	0.5		4.2	
BAP (5 mg L^{-1}) + KN(5 mg L^{-1})	1	0.8	4.8	1.4
BAP (5 mg L^{-1}) + KN (5 mg L^{-1}) + $AdSO_4$(50 mg L^{-1}) + AC (2 g L^{-1})	2.9	2.7	9.6	4.7
BAP (1 mg L^{-1}) + KN(1 mg L^{-1}) + $AdSO_4$(50 mg L^{-1}) + AC (2 g L^{-1}) + GA_3 (2 mg L^{-1})	1.1	1.2	5.1	1.3
BAP (1 mg L^{-1}) + KN(1 mg L^{-1}) + $AdSO_4$ (50 mg L^{-1}) + AC (2 g L^{-1})+IBA (1 mg L^{-1})	0.8 (callusing at base)		4.4	

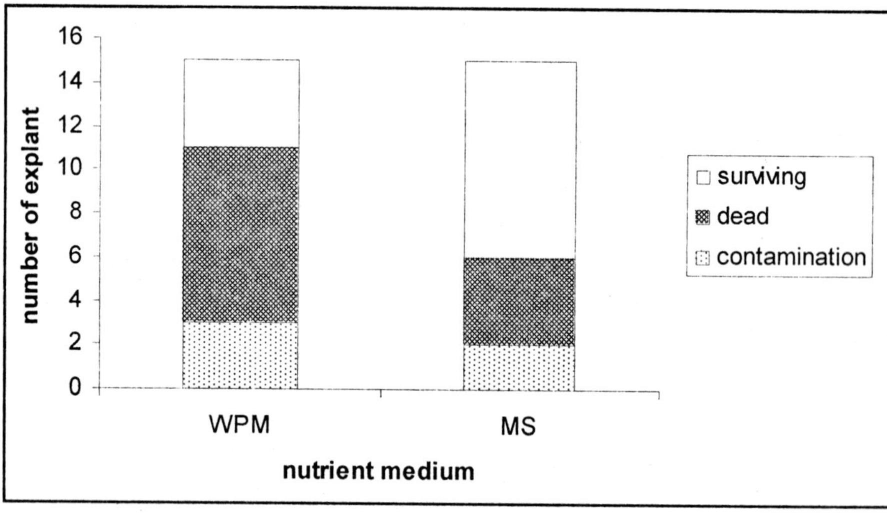

Fig. 3: Evaluation of the *in vitro* establishment of *S. Oleosa* after 4 weeks of culture Initiation.

The *in vitro* propagation of woody species poses problems due to high contamination rates and exudation of phenolics (Dalal *et al.*, 1992). Many solutions have been proposed to solve these problems without much success (Machado *et al.*, 1991). So far no reports on *in vitro* propagation of *Kusum* tree are available. In the present study we faced problem of contamination, browning, slow elongation of shoots and no rooting in cultures of *Kusum*. All these problems, except rooting, have been solved.

The treatment with 0.1% $HgCl_2$ was more effective than NaOCl in disinfection of explants of plum (Kumar *et al.*, 2002). $HgCl_2$ in different concentration and duration was reported most effective sterilant in several forestry trees viz., eucalypts, (Bisht *et al.*, 2000), mahua (Jagatram *et al.*, 2003), teak (Mascarenhas *et al.*, 1993) and sandal wood (Sanjaya *et al.*, 1997). In our experiment, $HgCl_2$ was more effective than NaOCl in surface sterilization of explants. Treatment of explants with Bavistin and agitation, helped in reduction of contamination. Bavistin solution without agitation was not effective in cashew culture as it could not reach the nodal cervices (Cardoza *et al.*, 1999). In rainy season, the contamination percentage was higher than in summer and winter seasons due to existence of high humidity during rainy season. Pretreatment of explants with the mixture of ascorbic (100 mg L^{-1}) and citric acid (150 mg L^{-1}) for 3 h and addition of this mixture in culture medium completely controlled exudation of phenolics (Kumar *et al.*, 2002). They also observed that addition of activated charcoal in the medium was less effective. Kant and Verma (1999) pretreated nodal explants of *Emblica officinalis* with chilled sterile antioxidant solution [AA (150 mg L^{-1}); AC (150 mg L^{-1}) and PVP (500 mg L^{-1})] for 1 h and also added these antioxidants (25 mg L^{-1}) into the medium to overcome the problem of leaching of phenolics.

Shoot initiation was observed in MS medium with BAP. However, for elongation of shoots Kinetin, $AdSO_4$ and AC played a major role. Vaidya and Bansal (2004) also achieved higher shoot length with the addition of $AdSO_4$ in culture medium. For several broadleaved tree species, thidiazurone was reported as an efficient substitute for BA in micropropagation. This cytokinin-like growth regulator can be tested for *S. oleosa* in future, when shoots have been multiplied in a larger scale. In spite of several trials, root initiation could not be achieved in *Kusum*.

CONCLUSION

Kusum tree seems to be a species that is recalcitrant for *in vitro* propagation. Effective methods to control contamination and browning of cultures were found out. This is the first successful report on shoot proliferation of this species in culture medium.

ACKNOWLEDGEMENTS

We thank Mr. R. Krishnamurthy (Director, Institute of Forest Productivity, Ranchi), Dr. Z. A. Haider (Head, Biotech. Div., Birsa Agriculture University, Ranchi) and Dr. K. G. Prasad (Ex-Director, Rain Forest Research Institute, Jorhat) for their constant encouragement during this study and to Mr. Atanu Sarkar and Mr. Sandeep Chakraborty for technical assistance. Financial support to this work was provided by the Department of Biotechnology, Government of India.

References

Bisht, P., Joshi, I., Chauhan, J. M. S. and Bagchi, S. K. (2000) Micropropagation of a 23 year old candidate plus tree of *Eucalyptus tereticornis* SM. *Indian J. For.* 23(2) : 149-154.

Cardoza, V., D'Souza, L. and D'Silva, I. (1999) Controlling contamination in cashew (*Anacardium occidentale* L.). In: *Plant Tissue Culture and Biotechnology Emerging Trends* (ed Kavi Kishor, P. B.) Universities Press, Hyderabad, pp.156-159.

Dalal, M. A., Sharma, B. B. and Rao, M. S. (1992) Studies on stock plant treatment and initiation culture mode in control of oxidative browning in *in vitro* cultures of grapevine. *Sci. Hort.* 51 : 35 – 41.

Das, N. and Chattopadhyay, R. N. (2003) Inventory of forest based medicinal plants-A case study in south West Bengal. *Indian For.* 129 (1) : 69-79.

Jagatram, M., Surendran, C., Paramathma, M., Prathiban, K. T. and Sasikumar, K. (2003) Micropropagation of M*adhuca latifolia. Indian J. For.* 26: 445-448.

Kant, U. and Verma, B. (1999) Propagation of *Emblica officinalis* Gaertn. through tissue culture. In: *Plant Tissue Culture and Biotechnology Emerging Trends* (ed Kavi Kishor, P. B.) Universities Press, Hyderabad, pp. 106-111.

Kumar, A., Vandana, Kumar, A. and Nautiyal, M. C. (2002) Micropropagation of Plum (*Prunus domestica* L.). In : *Trends in Plant Tissue Culture and Biotechnology* (eds Pareek, L. K. and Swarnkar, P. L.) Agrobios (India), Jodhpur, pp. 241-247.

Lloyd, G. and McCown, B. (1980). Commercially-feasible micropropagation of mountain laurel, *Kalima latifolia*, by use of shoot tip culture. *Pro. Intl. Plant Prop. Soc.* 30: 421-427.

Machado, M. L. C., Machado, A. C., Hanzer, V., Kalthoff, Weib, H., Mattanovich, D., Regner, F. and Katinger, H. (1991) A new, efficient method using 8-hydroxy-quinolinol-sulfate for the initiation and establishment of tissue cultures of apple from adult material. *Plant Cell Tissue Organ Cult.* 27 : 155-160.

Mascarenhas, A.F., Kendurkar, S.V. and Khuspe, S.S. (1993) Micropropagation of Teak. In: *Micropropagation of Woody plants* (ed Ahuja, M.R.) Kluwer Academic Publishers, Netherlands, pp. 247-262.

McGranahan, G. H., Driver, J. A. and Tulecke, W. (1987) Tissue culture of Juglans. In: *Cell and Tissue Culture in Forestry Vol.3* (eds Bonga, J. M. and Durzan, D. J.) Martinus Nijhoff Pub., Netherlands, pp. 261-271.

Murashige, T. and Skoog, F. (1962) A revised medium for rapid growth and bioassays with tobacco tissue culture. *Physiol. Plant.* 15:473-497.

Peltit, G.R., Numata, A., Cragg, G. M., Herald, D. L., Takada, T., Iwamoto, C., Riesen, R., Schmidt, J. M., Doubek, D. L. and Goswami, A. (2000) Isolation and structures of schleicherastatins 1-7 and schleicheols 1 and 2 from the teak forest medicinal tree *Schleichera oleosa. J. Nat. Prod.* 63 (1): 72-78.

Pradeepkumar, G. and Prathapasenan, G. (2003) Ethnobotany of Shool Paneshwar wild life sanctuary in Gujrat, A preliminary survey. *Indian For.* 129: 1322-1328.

Sanjaya, H. S., Ananthapadmanabha and Rai, V. R. (1997) *In vitro* shoot multiplication from the mature tree of *Santalum album* L. *MyForest* 33 (3) :45-49.

Sinha, A. (2005) Selection of plus trees of *Schleichera oleosa* (Lour.) Oken - A multipurpose tree. In : *Multipurpose Trees in the Tropics : Management and Improvement Strategies* (eds Tewari, V. P. and Srivastava, R. L.) AFRI, Jodhpur, pp. 719-721.

Vaidya, K. and Bansal, Y.K. (2004) Multiple shoot induction from apical buds of mature Semul (*Bombax ceiba* L.). *Indian J. For.* 27 : 175-178.

❑❑❑

Forest Biotechnology in India © 2008
Editors: S.A. Ansari, C. Narayanan & A.K. Mandal, pp.215-228
Satish Serial Publishing House, Delhi

21

Micropropagation Through Axillary Shoots Proliferation and Somatic Embryogenesis in Sandal Wood (*Santalum album* L) from Mature Trees

T.S. RATHORE, MAMATHA RANGASWAMY, BINIYA GOYAL, ABHAY KUMAR DUBEY AND P.S. RAO

Tree Improvement and Propagation Division, Institute of Wood Science and Technology, 18th Cross, Malleswaram, Bangalore-560 003, India

ABSTRACT

We developed *in vitro* clonal propagation protocols through axillary shoot proliferation and somatic embryogenesis in *S. album.* Nodal shoot segment proved to be the best for multiple shoot (4-6 shoots/explant) induction on MS medium supplemented with additives (ascorbic acid + citric acid + cysteine), NAA (0.1 mg/l) and BAP (0.1 - 2.5 mg/l), within 4 weeks at 25±1°C temperature and 2500 lux light. Up to 4 fold shoot multiplication was obtained on MS medium with additives, IAA and BAP under 2500 lux light at 12 h photoperiod. Sub-culturing within 5-6 weeks period on shoot multiplication medium proved ideal for maintenance of growth and vigour of shoot multiplication cultures. Four fold shoot multiplication rate was achieved with 5-6 weeks period on shoot multiplication medium. For rooting, combined use of IAA and IBA (1.25 mg/l) in the MS/4 medium proved better than IBA alone. Pulse treatment of shoot with IBA (1000-2500 ppm) was effective for *in vitro* and *ex vitro* rooting. Leaf, as an explant, proved to be the best for high frequency fragile callus on MS medium with additives [(ascorbic acid 50 mg/l + citric acid 25 mg/l cysteine 25 mg/l + glutamine (100 mg/l) + 2,4-D (1.0 mg/l)]. Synchronized Somatic embryogenesis (SE) induction was observed on MS medium with additives. MS medium with additives and ABA proved the best for synchronized development and maturation of SE's. MS/4 medium with IAA, GA_3 and activated charcoal proved the best for SE germination.

Keywords : Axillary shoot proliferation, *in vitro* and *ex vitro* rooting, Sandal wood, *Santalum album*, somatic embryogenesis.

INTRODUCTION

Santalum album L., commonly known as Indian sandalwood or chandan (family: Santalaceae), is an hemi-root parasite tree species. The genus *Santalum* consists of 16 species with predomination of *S. album* in India, Indonesia and Fiji. Sandalwood is commercially known as East Indian sandalwood and its fragrant heartwood produces an essential oil known as East Indian sandalwood oil. Indian sandalwood contains maximum oil (4 to 6%) and santalol (α and β santalols) contents, which are used in perfume fixative. The oil is also used in toiletries, cosmetics and pharmaceuticals (Shankaranarayan and Theagarajan, 2000) and also has excellent medicinal properties. *S. album* is mainly confined to peninsular India, but also found in other parts of the country. Karnataka and Tamil Nadu together contribute to more than 90% of natural sandalwood forests of the country. Bioresource of sandalwood has been over exploited for decades and are threatened in its habitats (Meera *et al.*, 2000). Furthermore illicit felling and smuggling have selectively eroded germplasm of the best trees of sandalwood from natural habitats. Now-a-days, it is difficult to locate trees more than 50 cm in girth. India has shown substantial decline in production of sandalwood from 3176 tons/year in 1960-61 to 1500 tons/year in 1997-98 and less than 1000 ton in 2000-01 (Shashidhar and Arunkumar, 1999; Ananthapadmanabha, 2000 a,b). Market trend indicated that sandalwood prices have substantially increased from Rs. 365/ton in the year 1900 to Rs. 6.5 lakhs/ton in 1999-2000. Current cost of sandalwood and oil are as Rs. 25 lakhs/ ton and Rs. 30,000/kg, respectively. Increase in costs is mainly due to heavy demand and drastic decrease in supplies.

Sandalwood was declared as a royal tree in 1792 by Tippu Sultan, then emperor of Mysore and the tree was considered as a government property (Srinivasan *et al.*, 1992). Cumbersome cultivation norm and royalty imposed by the government together discouraged farmers from its cultivation. Recent amendment in sandalwood cultivation policy by the Karnataka and Tamil Nadu state governments appears promising for the development of bioresource of sandal in the country and demand for the quality planting stock of sandal wood is increasing.

Sandalwood is predominantly an outbreeding species (Jyothi *et al.*, 1991; Bhaskar, 1992; Veerendra and Padmanabha, 1996) and exhibits variability in size and shape of leaf, flowering pattern, fruit/ seed size, heartwood and oil content (Srinivasan *et al.*, 1992; Srimathi *et al.*, 1995). Vegetative propagation of *S. album* has been reported through root suckers, root cutting, air layering, and grafting and stem cuttings of less than one - year old seedling (Srinivasan *et al.*, 1992). Above cited methods are inefficient for large scale production of clonal plants.Earlier studies on *in vitro* regeneration of *S. album* reported somatic embryogenesis from hypocotyl callus (Bapat and Rao, 1979), stem callus from shoot tip of 20-year-old plants through somatic embryogenesis (Lakshmi Sita *et al.*, 1979), cell suspension culture (Lakshmi Sita *et al.*, 1980a), regeneration of plantlet through somatic embryogenesis form protoplast (Rao and Ozias-Akins, 1985) and regeneration of triploid plant from endosperm (Lakshmi sita *et al.*, 1980b). Bapat and Rao (1988) reported production of synthetic seed by encapsuation of somatic embryos of *S. album.*

Success in shoot induction from mature trees through axillary shoot proliferation has been reported by Sanjay *et al.* (2001). Most of the earlier researches on *S. album* are based on explants from seedlings or young trees. None of the studies dealt with micropropagation of mature trees/plus trees/clones. In the present study, we carried out studies on critical factors influencing regeneration through axillary shoot proliferation and somatic embryogenesis using explants from natural trees.

MATERIAL AND METHODS

Management of source material and surface sterilisation

Selected plus trees and ramates of clones of sandal were periodically pruned to rejuvenate new shoots as a source of explants. Juvenile single nodal shoot segments and apical shoots obtained from pruned trees were used for *in vitro* regeneration through axillary shoot proliferation. Leaf and internodal segments were used for induction of somatic embryogenesis. Different explants, viz., nodal shoot segment, apical shoot, leaf and internode were washed in Tween 80 solution (0.1% v/

v) for 5 minutes, followed by washing with distilled water. Explants were also washed in systemic fungicide (Bavistin 0.1% w/v) solution for 5 minutes to minimize contamination. Explants were then surface sterilized in 70% ethanol for (15-30s) followed by three washes in sterile distilled water. Finally, explants were surface sterilized in 0.75 - 1.0 percent (w/v) solution of mercuric chloride for 4-5 minutes, depending on explant material, followed by 6 -7 washes with sterile distilled water.

Medium was solidified by addition of 0.6 percent agar-agar before autoclaving at 15 psi for 20 min. All inorganic salts of the medium were supplied by Hi-Media (Hi-Media, Bombay) whereas vitamins and growth hormones were supplied by Sigma (Sigma, USA). Depending on the requirement, culture tubes (25 x 150 mm), conical flasks (100 ml) and culture bottles (300 or 400 ml) were used. Shoot initiation, multiplication and rooting cultures were incubated at 25 $\pm$ 1°C temperature under 2500 lux light at 12 h photoperiod, whereas callus initiation and multiplication cultures were incubated under dark conditions.

Shoot initiation

Two types of explants, viz., apical shoot and nodal shoot segments, were tested to find their use as explant for high frequency multiple shoot initiation. In order to find out optimum requirement of growth hormones, auxins (IAA and NAA; 0.1 - 0.25 mg/l) and cytokinins kinetin (Kn) and BAP; 0.1 - 5.0 mg/l) were added either alone or in combination in the basal MS medium. To improve multiple shoot initiation frequency and reduce contamination problem, nodal shoot segments were collected during different seasons. Based on the initial results of the experiment on the effect of growth hormones on shoot initiation, MS medium with 0.1 mg/l NAA and 0.1 mg/l BAP was used to study the effect of period of collection on shoot initiation.

Shoot multiplication

In vitro induced shoots from the original explants either as a single shoot or clump of 2-3 shoots were used for further multiplication. Original explant was also repeatedly subcultured on the MS medium with 0.1 mg/l NAA and 0.1 mg/l BAP for four cycles to obtain new shoots for further shoot multiplication and rooting.

In order to optimize growth hormones and their concentrations, auxins (IAA and NAA) and cytokinins (kinetin and BAP) were tested using 0.1 mg/l -5.0 mg/l levels in MS medium. In a separate experiment, additives viz., ascorbic acid, citric acid, cysteine, glutamine, yeast extract, malt extract and coconut milk were also added in MS medium along with IAA (0.1 mg/l) and BAP (0.1 mg/l). To standardize exogenous requirement of energy in the form of sucrose, various concentration of sucrose, viz., 6.0 per cent (w/v) were added in MS medium along with IAA (0.1 mg/l) and BAP (0.1 mg/l). Shoot multiplication cultures were incubated at 25 $\pm$ 1°C temperature under 2500 lux light at 12 h photoperiod and sub-culturing was carried out within 5-6 weeks period to maintain growth phase.

Root initiation

In vitro shoots, produced either through repeated sub-culturing of original explants or shoots multiplied by sub-culturing on shoot multiplication medium, were tested for rooting. Two separate experiments were carried out on *in vitro* rooting. In the first experiment, various auxins viz., IAA, IBA, NAA and NOA, were added either alone or in combinations in MS/4 basal salts medium. In the second experiment various media, viz., WP, HE, White, various modifications of MS (MS/2, MS/4, MS/8 and MS/16), Hogland solution and Knop's solutions were used with IBA either alone or with IAA. Attempts were also made to test effect of pulse treatment with IBA (500-2500 ppm) followed by the transfer of shoots to hormone free medium, for *in vitro* rooting. For *ex vitro* rooting, shoots were pulse (5 min and 30 min) treated with various auxins (IAA, IBA, NAA and NOA 1000 ppm) followed by transfer to rooting medium (soilrite) in trays.

Callus initiation and multiplication

Leaf and internode were used as explant for callus initiation on MS medium with various auxins, viz., IAA, NAA, NOA and 2,4-D (1.0 - 2.0 mg/l) either alone or with Kn/BAP. Callus initiation cultures were incubated at 20 or 25°C temperature under dark conditions. Callus grown on MS medium with 2,4-D was sub-cultured on MS medium with various concentrations of 2,4-D (0.5 - 2.0 mg/l) either alone or in combination with BAP (0.1 mg/l) in order to standardize optimum requirement of 2, 4-D for further multiplication of embryogenic callus. Sub-culturing of callus was carried out within 4 weeks period on fresh callus multiplication medium.

To improve texture and quality of callus, two experiments were conducted on effect of additives, viz., ascorbic acid (50 mg/l), citric acid (25 mg/l), cysteine (25 mg/l), glutamine (100 mg/l), yeast extract (500 mg/l), malt extract (500 mg/l) and coconut milk (10% v/v) in MS medium along with 2,4-D 1.0 mg/l. Another experiment was conducted on different incubation condition, viz., 20, 25 and 30°C under dark and 25°C under light, to optimize conditions to maintain embryogenic callus for a long time.

Somatic embryo initiation

Embryogenic callus grown on MS medium with various additives (ascorbic acid, citric acid, cysteine, glutamine and 2,4-D) was used for somatic embryogenesis. In order to induce high frequency somatic embryogenesis with synchronized development, studies were conducted on the effect of PGR, additives and sucrose. MS medium was used with various growth hormones, viz., IAA, IBA, NAA, NOA and 2,4-D (1.0 mg/l) with and without BAP (1.0 mg/l). In a separate experiment various additives, viz., ascorbic acid, citric acid, cysteine, glutamine, yeast extract, and malt extract were used either alone or in combination, to improve frequency and synchronized development of embryos from the embryogenic cultures. To improve size of somatic embryo, various concentrations of sucrose viz., 1.5 – 6.0 percent (w/v) were tested in MS medium with additives (ascorbic acid, citric acid, cysteine and glutamine), on hormone free medium.

Somatic embryo maturation

In order to achieve synchronized development and maturation of somatic embryos, growth hormones, viz., GA_3 and ABA, alone and in combination, as well as without growth hormones, were tested in MS medium. In another experiment, two concentrations of PEG (3.75% and 7.5%) were tested alone or with ABA (0.5 mg/l and 1.0 mg/l) in the MS medium with additives.

Somatic embryo germination

Late torpedo and bipolar embryos were used for germination studies. Five treatment viz., MS/4 hormone free, MS/4 with GA_3, MS/4 with IAA, MS/4 with IAA and GA3 and MS/4 with IAA, GA_3 and activated charcoal were tested for germination of somatic embryos. In another experiment, various media, viz., MS, MS/4, WP and HE supplemented with IAA and GA_3 were tested to achieve high frequency germination of somatic embryos.

RESULTS

Shoot initiation

Nodal shoot segment was found better than apical shoot segment for multiple shoot initiation. Apical shoot segment was very sensitive to sterilants and exhibited very slow response to shoot induction and subsequent growth. Therefore, nodal shoot segment was used for the various experiments on shoot initiation.

Among the various treatments of growth hormones, MS medium added with NAA (0.1 mg/l) and BAP (1.0 - 2.5 mg/l), proved ideal for high frequency multiple shoot (4-6 shoots/segments) initiation on MS medium at 25 $\pm$ 1°C temperature under 2500 lux light at 12h photoperiod within 4 weeks period. Higher concentration of BAP produced dwarf shoots. Medium with NAA and Kn proved less effective in all the concentrations than with NAA and BAP. Lower concentrations (0.1 and 0.5 mg/l) of BAP in the medium were not effective in induction of multiple shoots.

Explants collected during monsoon period (June to September) and winter (December to January) posed problem of bacterial contamination as well as poor response to shoot induction. Explants collected during February to May from the pruned trees proved the best for the establishment of high frequency shoot initiation cultures.

Shoot multiplication

Repeated sub-culturing of original shoot segment produced shoots up to 4th cycle. From a single explant, 40-50 shoots were obtained within three months. Such isolated 2-3 shoots/clump can be sub-cultured on fresh medium for further multiplication. MS medium with 0.1 mg/l NAA and 1.0 mg/l BAP proved ideal for repeated sub-culturing of original explants with shoots.

Among the various growth hormones NAA (0.1 mg/l) and BAP (1.0 mg/l) proved the best combination for high frequency multiplication. Multiplication rate of 3-4 fold within 5 weeks period at 25 $\pm$ 1°C temperature under 2500 lux light at 12 h photoperiod, was achieved. Further increase in BAP concentration in the medium resulted in drastic reduction in shoots size.

Among the various media tested, MS medium supplemented with ascorbic acid (50 mg/l) and citric acid (25 mg/l) and cysteine (25 mg/l), proved to be ideal for shoot multiplication and subsequent growth. Shoot produced on this medium were better than medium without additives. However, incorporation of additives in the medium minimized leaching and browning problems.

Out of various concentrations of sucrose tested, 3.0 and 4.5% sucrose levels proved better for shoot multiplication and growth. Poor shoot growth and multiplication rate was observed on very low (1.5%) or very high (6.0%) sucrose levels in the shoot multiplication medium (MS, additives, 0.1 mg/l NAA and 1.0 mg/l BAP).

Sub-culturing of shoots within 6 weeks was essential to maintain growth phase and multiplication rate. Shoot clump proved better than single shoot or shoot segment for further multiplication of shoots. Five inoculum of 2-3 shoot/clump produced 80-100 shoots in 300 ml culture bottles within 8-10 weeks period on shoot multiplication medium. About 80% of shoot can be used for the root initiation and remaining 20% shoots, which are comparatively smaller in length, can be used for the further multiplication. Shoot cultures were maintained for more than two years without any sign of deterioration or loss of multiplication capability.

In vitro root initiation

Among the various auxins tested, combined use of IAA with IBA in MS/4 basal salts medium proved the best, followed by same medium with IBA alone. Medium with NAA or NOA in the medium also induced roots, but callus from the shoot base was also encountered. IAA alone in the medium proved less effective. Initial incubation of cultures for one week under dark, followed by incubation under light condition proved effective for early root initiation. Pulse treatment of shoot with IBA (1000/2500 ppm) and subsequent transfer of shoots to hormone free MS/4 basal medium was better than other concentrations of IBA.

Somatic embryogenesis: Callus initiation and multiplication

Among the various types of explants used for callus initiation, leaf was found to be the best choice followed by internode. Out of the auxins used, 2, 4-D (1.0 mg/l) proved to be the best for achieving high rate (100%) of callus induction from the leaf explants. IAA and IBA, either alone or with Kn/BAP in the medium did not induce callus. Very low frequency (30-40%) of callus induction was observed with NAA and NOA either alone or with Kn/BAP.

Callus grown on the MS medium with 2,4-D (1.0 mg/l) under dark condition at 20 $\pm$ 1°C temperature performed better. High concentration of 2,4-D in the medium did not improve quality/intensity of callus. Incorporation of BAP (0.1 mg/l) in the medium was ineffective in improvement of callus multiplication. Addition of additives, viz., ascorbic acid (50 mg/l), citric acid (25 mg/l), cysteine (25 mg/l) and glutamine (100 mg/l) improved texture and quality (more granular and healthy) of callus in MS medium with 2,4-D (1.0 mg/l).

Among the four incubation conditions callus culture incubated in MS medium at 20 and 25°C under dark favoured multiplication and maintenance of the embryogenic callus.

Somatic embryo initiation

Among the various growth hormones MS IAA (1.0 mg/l), followed by IBA (1.0 mg/l) induced high frequency (100%) synchronized somatic embryos without compaction. NAA and NOA induced somatic embryos but did not favour synchronization, whereas 2,4-D induced only callus. MS medium with IAA (1.0 mg/l) and BAP (1.0 mg/l) induced somatic embryos but synchronization was poor and embryos fused due to compaction. Incorporation of additives improved intensity, quality and synchronization of the embryos. Addition of yeast or malt extract improved intensity of embryo induction, but did not improve synchronization.

Somatic embryo development and maturation

MS medium with additives and ABA proved best for synchronized development and maturation of somatic embryos. Medium without ABA also favoured development of somatic embryos but synchronization was not compared to medium containing ABA. Medium GA_3, alone or with ABA, was not found suitable for somatic embryo development. Even under diffuse light/dark condition, embryos were greenish in colour and leafy structures developed.

Improvement in synchronized development and maturation was observed with the addition of PEG in the medium. Combined use of ABA (1.0 mg/l) and PEG (3.75%) proved best for synchronized development and maturation of the embryos. On this medium embryos matured to late - torpedo and bipolar stages, and were whitish in colour and there was no development of secondary embryos.

Germination of somatic embryos

Among the five treatments, MS/4 medium with IAA (1.0 mg/l), GA_3 (1.0 mg/l) and activated charcoal proved the best for germination of somatic embryos. Incubation of somatic embryos on the above medium for two weeks under dark conditions supported early and maximum germination. Medium with IAA favoured maximum root development. GA_3 (1.0 mg/l) favoured shoot development and elongation. Heller medium favoured maximum (90%) germination. This medium induced early and high frequency rooting, but shoot were rudimentary and growth was also very slow. MS/4 medium not only proved best for overall recovery of plantlets but also favoured shoot as well as root development.

Plantlets raised through axillary shoot proliferation and somatic embryogenesis were hardened for 4-5 weeks in culture bottles or plastic trays with soilrite before transplantation to potting mixture. Hardening was done for two months in mist chamber/green house. While plants were morphologically normal and similar to parent plants their genetic fidelity need to be tested before large scale production.

DISCUSSION

Management of source (ortet /mother) tree through lopping, pruning or pollarding is known to improve juvenility by ensuring new shoots and minimizing contamination. In the present study, explants obtained from the pruned tree proved better than those from unpruned trees, for establishment of cultures. Shekhawat *et al.* (1993) reported effect of lopping and pruning on the availability of juvenile material for a longer period of time in *Prosopis cineraria*. In *S. album,* nodal shoot segment proved better than apical shoots for the production of multiple shoots. Similar results have been reported in *Tecomella undulata* (Rathore *et al.,* 1991). Combined use of NAA and BAP (1.0 - 2.5 mg/l) in the MS medium supplemented with additives (ascorbic acid, citric acid and cysteine) favoured high rate of multiple shoot (4-5 shoots/explants) induction. Kinetin, alone or with NAA, was not effective as compared to BAP and NAA.

Explant collected during March - May proved excellent for culture establishment and induction of multiple shoots, without much contamination. In *Prosopis cineraria,* Shekhawat *et al.* (1993) observed significant effect of season (period) on explants during shoot initiation. Repeated sub-culturing of original explant could be continued up to four cycles for production of shoots. This approach provided scope for development of shoot clumps, which can earlier be used for further multiplication.

Shoot clump (2-3 shoots/clump) proved better than other explants for rapid multiplication and further growth. MS medium, fortified with additives, NAA/IAA (0.1 mg/l) and BAP (1.0 mg/l) proved suitable for further multiplication of differentiated shoots. In a single culture bottle, about 80-100 shoots could be produced within 8-10 weeks period, which can be used for rooting. Sub-culturing of shoots within 5-6 weeks on fresh shoot multiplication medium maintained growth phase and vigour up to two years. Four-fold multiplication was obtained within 5-6 weeks period, on shoot multiplication medium.

Results of *in vitro* rooting experiment revealed that combined application of IAA and IBA in the MS/4 basal salts medium supported maximum roots. Among auxins, IBA was found to be the best for rooting. Superiority of IBA over the other auxin source has also been reported in various other tree species (Rathore *et al.*, 1991; Shekhawat *et al.*, 1993).

Explant type, source, genotype, medium and growth hormones are considered important factors, for induction of embryogenic callus. In the present study, leaf explant produced soft, fragile and granular callus on MS medium with additives and 2, 4-D. Other auxins, viz., IAA, IBA, NAA and NOA were ineffective in induction of embryogenic callus. Kato (1996) reported development of embryogenic callus from immature leaf on medium containing 2,4-D. In *S. album* embryogenic callus could be maintained for a long period by sub-culturing in 4-5 weeks cycles to fresh callus multiplication medium.Incubation condition influenced multiplication and quality of callus cultures. Cultures incubated at 20 and 25°C under light and at 30°C in dark resulted in compaction of callus and gradual browning of callus cultures.

Medium and growth hormones play important role in somatic embryo induction and development (Huang *et al.*, 1998). In the present study, MS medium, supplemented with additives and IAA, proved the best medium for synchronized somatic embryo induction and development. Incorporation of BAP with IAA in the same medium resulted in low intensity embryo induction, more compaction, poor synchronization and secondary embryo development.

In *S. album,* incubation of culture at 20 and 25°C in dark proved suitable for synchronized embryo induction. Higher temperature (30°C) in dark and 25°C under light resulted in low intensity, unsynchronized development and compactness in the embryo cultures. In contrast, Takamura *et al.* (1995) reported enhanced somatic embryogenesis at higher temperature (25 to 30°C).

High frequency synchronized development and their further conversion into bipolar embryo stage with well developed cotyledon are essential for high rate of germination of somatic embryos. Nutrient medium, ABA, PEG, carbohydrate and polyamines play a vital role in somatic embryo development and their maturation (Fuentes *et al.*, 2000; Pullman *et al.*, 2003; Lipavska *et al.*, 2004). In this study, combined use of PEG and ABA proved better for development and maturation of somatic embryos.

Success of somatic embryogenesis ultimately depends on germination and recovery of plantlets. Low nutrient medium, IAA, GA_3, low concentration of Kn or BAP are known to help in germination. In the present study, MS/4 basal medium with IAA and GA_3 proved the best for germination of somatic embryo into shoot and root.Although more than 90 percent germination could be obtained on Heller medium with IAA and GA_3, further shoot growth was effected.

CONCLUSION

Micropropagation through axillary shoot proliferation and somatic embryogenesis have high potential for rapid and mass production of clonal planting material from plus trees/clones/elite genotypes of *S. album.* Hardening is the most crucial step in establishment of sandal plants, which require compatible hosts for their survival. Assessing genetic fidelity of the plants produced through axillary shoot proliferation and somatic embryogenesis using DNA markers will be essential before large scale operational development of clonal planting material.

ACKNOWLEDGEMENT

This research work was carried out with the financial assistance of DBT, Government of India, New Delhi. Director and Group Coordinator (Research) have provided facilities and constant encouragement to carryout research work.

References

Ananthapadmanabha, H. S. (2000a) The present status of sandalwood in India and Australia. *Fafai Journal* 2 (1) : 33-36.

Ananthapadmanabha, H. S. (2000b) Sandalwood and its marketing trend. *My Forest* 36 (2) : 147-151.

Bapat, V. A. and Rao, P. S. (1979) Somatic embryogenesis and plantlet formation in tissue culture of Sandalwood *(Santalum album* L.). *Ann. Bot.* 44 : 629-630.

Bapat, V. A. and Rao, P. S. (1988) Sandalwood plantlet from synthetic seeds. *Plant Cell Rep.* 7 : 434-436.

Bhaskar, V. (1992) Pollination biology and fertilization in *Santalum album* L. (Santalaceae). *Flora (Jena)* 187: 73-78.

Fuentes, S. R. L., Caheiros, M. B. P., Hanetti - Filho, J. and Vieira, L. G. E. (2000) The effect of silver nitrate and different carbohydrate sources on somatic embryogenesis in *Coffea canephora Plant Cell Tissue Organ Cult.* 60: 5-13.

Huang, F. H. and Gbur, Jr. E. E. (1998) Effect of basal medium, growth regulators and their concentration on initiation of embryogenic cultures from immature zygotic embryos of loblolly pine (*Pinus taeda* L). *Plant Cell Rep.* 17: 298-301.

Jyothi, P. V., Atluri, J. B. and Subba Reddi, C. (1991) Pollination ecology of *Santalum album* (Santalaceae). *Trop. Ecol.* 32: 98-104

Kato, M. (1996) Somatic embryogenesis form immature leaf of *in vitro* grown tea shoots *Plant Cell Rep.* 15: 920-923.

Lakshmi Sita, G., Raghava Ram, N. V. and Vaidyanathan, C. S. (1979) Differentiation of embryoids and plantlets from shoot callus of sandalwood. *Plant Sci. Lett.* 15: 265-270.

Lakshmi Sita, G., Raghava Ram, N. V. and Vaidyanathan, C. S. (1980a) Regeneration of whole plants by embryogenesis from cell suspension cultures of sandalwood. *Curr. Sci.* 49:196-198.

Lakshmi Sita, G., Raghava Ram, N. V. and Vaidyanathan, C. S. (1980b) Triploid plants from endosperm cultures of sandalwood by experimental embryogenesis. *Plant Sci. Lett.* 20: 63-69.

Lipavska, H. and Konradova, H. (2004) Somatic embryogenesis in confers: The role of carbohydrate metabolism. *In Vitro Cell. Dev. Biol. – Plant* 40: 23 – 30.

Meera, C., Nageswara Rao, M., Ganeshaiah, K. N., Uma Shaankar, R. and Swaminath, M.H. (2000) Conservation of sandal genetic resources in India: 1. Extraction patterns and threats to sandal resources in Karnataka. *MyForest* 36 (2): 125-132.

Pullman, G. S., Namjoshi, K. and Zhang, Y. (2003) Somatic embryogenesis in loblly pine (*Pinus taeda* L): improving culture initiation with ABA and AgNo3. *Plant Cell Rep.* 22 : 85-95.

Rao, P. S. and Ozias - Akins, P. (1985) Plant regeneration through somatic embryogenesis in protoplast cultures of sandalwood. *Protoplasma* 124: 80-86.

Rathore, T. S., Singh, R. P. and Shekhawat, N. S. (1991) Clonal propagation of desert teak (*Tecomella undulata*) through tissue culture. *Plant Sci.* 79: 217-222.

Sanjay, Rathore, T. S. and Rai, R. V. (2001) *In vitro* cloning of Indian Sandal (*Santalum album* L.) a highly prized over - exploited tree species. In: Biotechnological Innovation in conservation and analysis of plant diversity, pp. 63.

Shankaranarayana, K. H. and Theagarajan, K. S. (2000) Recent development in chemistry and utilization of sandalwood. *Wood News* 10 (1): 17-20.

Shashidhar, K. S. and Arun Kumar, A. N. (1999) Trade in Sandal (*Santalum album* L.) *Fafai Journal* 1 (4): 51-56.

Shekhawat, N. S., Rathore, T. S., Singh, R. P., Deora, N. S. and Rao, S. R. (1993) Factors affecting *in vitro* cloning of *Prosopis cineraria. Plant Growth Regul.* 12: 273-280.

Srimathi, R. A., Kulkarni, H. D. and Venkatesan, K. R. (1995) *Recent Advances in Research and Management of Sandal (Santalum album L.) in India.* Associated Publishing Co., New Delhi.

Srinivasan, V. V., Sivaramakrishnan, V. R., Rangaswamy, C. R., Ananthapadmanabha, H. S. and. Shankaranarayana, K. H. (1992) *Sandal* (*Santalum album* L.). Institute of Wood Science and Technology (ICFRE), Bangalore.

Takamura, T., Miyjima, I. and Matsuo, E. (1995) Somatic embryogenesis of *Cyclamen persium* Mill 'Anneke' from aseptic seedlings. *Plant Cell Rep.* 15: 22-25

Veerendra, H. C. S. and Padmanabha, H. S. A. (1996) The breeding system in sandal (*Santalum album L.*) *Silvae Genet.* 45: 188-190.

❑❑❑

Forest Biotechnology in India © 2008
Editors: S.A. Ansari, C. Narayanan & A.K. Mandal, pp.229-236
Satish Serial Publishing House, Delhi

22

Micropropagation of Acacia Hybrid (*Acacia mangium* x *A. auriculiformis*)

R. SUMATHI, P. MALLIGA, K.S. RATHNAM
R. YASODHA AND K. GURUMURTHI

Division of Plant Biotechnology, Institute of Forest Genetics and Tree Breeding, Coimbatore-641 002, India

ABSTRACT

The acacias, the fast growing, short rotation and nitrogen fixing tree species, are attracting more attention as an important source of pulpwood. Clonal multiplication of inter-specific hybrid (*Acacia mangium* x *A. auriculiformis*) is a widely used strategy to multiply desirable genotypes. In the present study, large-scale clonal propagation was achieved through axillary bud proliferation, using explants collected from 4-year-old interspecific hybrids. Axillary bud cultures were established in MS medium supplemented with 4.44 mM BAP. The bud proliferation was achieved in MS medium with 2.22 mM BAP. Even after 20-22 sub-culturing in the proliferation medium, the shoot cultures restored the seedling trait, i.e. the pinnate leaves. Shoots (4-6 cm) were used for rooting. *In vitro* rooting response in half strength MS medium with NAA ranged from 67-94 percent depending upon the clone. Under *ex vitro* conditions, 40-50 percent rooting was achieved when shoots were treated with 100 ppm IBA solution. Rooted plantlets were successfully hardened and planted in field.

Keywords : Ac climatization, axillary bud, *in vitro* / *ex vitro* rooting, shoot multipalication.

INTRODUCTION

The genus *Acacia* includes about 1200 species, most of which are important sources of pulpwood, firewood, fodder and tannin. They are fast growing, nitrogen fixing tree species with short rotation, suitable for reforestation programme in the tropics. They also yield quality timber, and hence, their plantations are being raised (Anonymous, 1989) on a large scale. In establishment of hard wood plantation, micro propagation is preferred for large scale production of quality planting material. Micropropagation is a rapid method for producing clonal planting stock of trees for afforestation, production of woody biomass and conservation of elite germplasm (Bonga and Durzan, 1982). To enhance *in vitro* response in mature trees, various *ex vitro* treatments, prior to culturing, have been recommended (von Aderkas and Bonga, 2000). Juvenile tissues from mature tree have been used for large-scale propagation in order to overcome problem of inadequate adventitious root formation and found to be highly responsive for *in vitro* regeneration. The paper presents results of preliminary studies on clonal multiplication of axillary buds of one-year-old clones of *Acacia* hybrid (*A. mangium* x *A. auriculiformis*).

MATERIALS AND METHODS

Plant material

Axillary buds collected from 4-year-old fast growing acacia hybrid clones, viz., SU-3, SU-40 and HT-7 were used for *in vitro* shoot multiplication.

Shoot multiplication

One day prior to the collection of buds, the clones were sprayed with 0.1% Bavistin and 0.01% Streptomycin. After collection, the bud explants were washed thoroughly in distilled water, treated with 0.1 % Savlon® for 5 minutes, disinfected with 0.1 % Bavistin and 0.1 % Streptomycin for 20 minutes and 0.1 % Mercuric chloride for 5 minutes, and finally rinsed three times with sterile distilled water. The explants were then placed on MS (Murashige and Skoog, 1962) medium added with 4.44-μM 6-benzyl adenine (BAP), 0.6% agar and 2% sucrose (pH 5.6). The cultures were maintained at 25 ± 2°C under 16 hours photoperiod. After 55 days of maintenance on medium, the buds showed stunted growth and yellowing of phyllode tips. Therefore, the buds were transferred to MS medium supplemented with 2.22 μM

BAP for shoot proliferation. Fresh *in vitro* shoots were regularly sub-cultured at 4-5 weeks interval.

In vitro rooting and acclimatization

For *in vitro* root induction, excised micro shoots (4 - 6 cm length) were transferred to solidified MS medium supplemented with indole 3-butyric acid (IBA), activated charcoal and various concentrations of naphthyl acetic acid (NAA). All cultures were incubated at 25 ± 2°C under 16 hours photoperiod.

For *ex vitro* rooting, cut ends of 4-6 cm long micro cuttings were dipped in 100 ppm IBA solution for 5 minutes, transferred to root trainers filled with presoaked vermiculite, and maintained inside poly tunnels. After 30 days, data on rooting was recorded.

RESULTS AND DISCUSSION

Culture establishment

After 8-10 days, the axillary buds produced adventitious shoots in the initiation medium (Fig. 1a). Multiple shoot formation appeared to be controlled by the growth substances and their concentration. It was observed that in *A. mangium,* shoot multiplication was maximum in MS medium with 0.5 mg BAP; further increase in BAP concentration decreased the shoot multiplication. In the present study, the explants produced multiple shoots after 3rd subculture in multiplication medium. Micro shoots produced pinnate leaf after 4th subculture (Fig. 1b); the phyllodes were narrow and pointed, resembling phyllodes of seedlings. The expression of such seedling traits in tissue culture is significant since it indicated juvenility. Leaf is a readily identifiable morphological marker of juvenility. Some biochemical markers like peroxidase and esterase isozymes (Brand and Lineberger, 1992; Huang *et al.*, 1996), protein phosphorylation (Kuo *et al.*, 1995), ethylene production (Huang *et al.*, 2000), polyamine concentration (Rey *et al.*, 1994), poly phenols (Fernandez-Lorenzo *et al.*, 1999) and photosynthetic potential (Huang *et al.*, 2003) have been employed to assess juvenility of plant tissue. However, such biochemical markers have limited utility due to inadequate experimental models.

Shoot multiplication, *in vitro* and *ex vitro* rooting

After 10th subculture, average number of shoots per culture flask was 8 and the number of harvestable shoots (length = 3 cm) was 2-3 per flask. After 20th subculture, average number of shoots increased to 9-12 and harvestable shoots to 5-8 per flask. Number of shoots, shoot length and harvestable shoots per flask were maximum in SU-3 followed by SU-40 and HT-7 (Table1). Sub-culturing the shoots several times increased multiplication rate (Fig 1c) and rooting. *In vitro* rejuvenation of mature tissues was essential before the micro-shoots could be induced to root. In redwood *in vitro* shoots developed from 8-year-old tree required 11 subcultures for rejuvenation while shoots from 25-year-old trees required 20 or more subcultures (Ahuja, 1996).

Table 1: Multiplication rate of *Acacia* hybrid (*A. mangium* x *A. auriculiformis*) clones after 10th and 20th subculture.

Subculture	Clone	No of shoots per flask	Average shoot height	No of shoots with pinnate leaf	Harvestable shoots per flask
10th subculture	HT-7	8.6 ± 0.79	2.94 ± 0.25	0.8 ± 0.2	2.6 ± 0.83
	SU-3	8.4 ± 0.56	3.12 ± 0.19	2.0 ± 0.37	3.1 ± 0.80
	SU-40	8.4 ± 0.91	2.44 ± 0.11	1.8 ± 0.61	1.5 ± 0.37
20th subculture	HT-7	9.9 ± 0.78	2.60 ± 0.12	2.5 ± 0.31	2.5 ± 0.68
	SU-3	11.4 ± 0.65	3.42 ± 0.20	5.0 ± 0.76	5.0 ± 0.83
	SU-40	12.7 ± 0.91	2.95 ± 0.13	4.9 ± 0.38	4.5 ± 0.48

After repeated sub-culturing and subsequent transfer to rooting medium, shoot with length more than 3 cm developed roots after 20 days (Fig. 1d). SU-3 showed maximum (89 %) rooting in MS media with 32.2 mM NAA followed by SU-40 (67 %) in MS media with 43 mM NAA and HT-7 (94 %) in MS media with 53.7 mM NAA (Table 2). MS with 24.6 mM NAA was found to be the best medium for all the three clones since rooting ranged 63-79 %. Under *ex vitro* condition, 40-50 % rooting was achieved. In a hybrid of *E. nitens* and *E. grandis*, up to 80% rooting was obtained after *in vitro* rejuvenation (Mokotedi, 1999). Rooted plants were transplanted in root trainers filled with vermiculite under shade, and acclimatized inside the poly tunnels. After two months, both *in vitro* and *ex vitro* rooted plants (Fig. 1e) transferred to polybags filled with potting mixture (Sand, soil and manure) showed survival rate of 90-100 %.

In acacias, different types of explants like coppice shoot in *A. auriculiformis* x *A. mangium* (Banik and Islam, 1996) and *A. mearnsii* (Beck *et al.*, 1998, Beck and Dunlop, 1999), apical meristems from

Fig 1. Micropropagation of *Acacia* Hybrid (*A. mangium* X *A. auriculiformis*)
a. Establishment of axillary bud in culture. b. 70-day-old axillary bud with new shoots and pinnate leave. c. Multiple shoot production. *d. In vitro* rooted plants in root trainers. *e. Ex vitro* rooted plants.

Table 2: Rooting response of micro shoots of *Acacia* hybrid (*A. mangium* x *A. auriculiformis*)

Media composition	Clone	% rooting	No of roots	Root length (cm)	Shoot height (cm)	No of leaves
¾ strength MS + 2 mg/l activated charcoal	SU – 3	20	1.20 ± 0.20	4.05 ± 0.76	4.74 ± 0.68	2.80 ± 0.49
	SU – 40	14	1.00 ± 0.00	3.60 ± 0.10	3.75 ± 0.75	2.50 ± 0.50
	HT – 7	22	1.00 ± 0.00	5.40 ± 2.06	4.83 ± 0.73	4.33 ± 0.33
Half strength MS + 24.6 µM IBAP + 26.8µM NAA	SU – 3	63	7.78 ± 0.66	1.81 ± 0.24	3.97 ± 0.33	1.89 ± 0.45
	SU – 40	64	6.13 ± 1.36	3.19 ± 0.67	3.03 ± 0.34	1.50 ± 0.19
	HT – 7	79	3.30 ± 0.83	3.07 ± 0.30	4.00 ± 0.18	3.00 ± 0.30
Half strength MS + 32.2 µM NAA	SU – 3	89	3.10 ± 0.53	4.37 ± 0.44	3.55 ± 0.33	2.80 ± 0.36
	SU – 40	47	4.10 ± 0.71	4.96 ± 0.85	3.85 ± 0.45	2.90 ± 0.34
	HT – 7	65	2.70 ± 0.33	4.93 ± 0.39	4.62 ± 0.34	2.80 ± 0.25
Half strength MS + 43.0 µM NAA	SU – 3	79	3.70 ± 0.45	3.34 ± 0.13	3.28 ± 0.26	2.60 ±0.22
	SU – 40	67	3.40 ± 0.73	3.86 ± 0.35	3.45 ± 0.27	2.70 ± 0.15
	HT – 7	68	1.80 ± 0.35	4.60 ± 0.52	3.02 ± 0.38	3.20 ± 0.44
Half strength MS + 53.7 µM NAA	SU – 3	57	4.10 ±0.66	3.60 ± 0.31	3.57 ± 0.33	2.80 ± 0.36
	SU – 40	63	3.11 ± 0.48	4.42 ± 0.66	3.34 ± 0.27	2.33 ± 0.17
	HT – 7	94	2.80 ± 0.63	5.00 ± 0.73	4.19 ± 0.37	3.20 ± 0.39
Knop's with out growth regulator	SU – 3	44	1.30 ± 0.15	2.80 ± 0.60	3.70 ± 0.21	2.20 ± 0.15
	SU – 40	14	1.00 ± 0.00	5.50 ± 0.61	2.30 ± 0.44	1.60 ± 0.24
	HT – 7	0				

micropropagated plantlets and mature trees of *A. mearnsii* (Beck *et al.*, 2000), axillary buds from seedlings of hybrid acacia (Darus, 1992), *A. mangium* (Darus, 1991), *A. stenophylla* (Crawford and Hartney, 1987), *A. seyall* (Capuli and Calinawan, 1999), *A. mearnsii* (Beck *et al.*, 1998, Marguerite *et al.*, 2001), *Acacia catechu* (Kaur and Kant, 2000) and *A. auriculiformis* (Ide *et al.*, 1994, Watanabe *et al.*, 1994), were used for inducing juvenility through micropropagation. The present study demonstrated that juvenility could be induced in *Acacia* hybrid by using axillary buds from hedged trees thereby restoring the seedling trait and enhancing the rooting potential.

References

Ahuja, M.R. (1996) Micropropagation and field testing of "frost tolerantly" Sequoia sempervirens genotypes. In: Proceedings of the conference on coast redwood forest ecology and management, Arcata, USA, pp. 153-155.

Anon (1989) *The Wealth of India (Vol. A-B) : Raw Materials.* Council of Scienti?c Industrial Research, New Delhi.

Banik, R. L. and Islam S. A. M. N. (1996) In vitro clonal propagation of hybrid acacia (*A. auriculiformis x A. mangium*). *Bangladesh J. For. Sci.* 25 : 1-7.

Beck, S. L. and Dunlop R. (1999) Vegetative propagation of the black wattle (*Acacia mearnsii* de Willd), *ICFRE Bull.* 8 : 15.

Beck, S.L., Dunlop R.. and van Staden J. (1998) Rejuvenation and micropropagation of adult *Acacia mearnsii* using coppice material. *Plant Growth Regul.* 26:149-153.

Beck, S. L., R. Dunlop. And J. van Staden. (2000) Meristem culture of *Acacia mearnsii. Plant Growth Regul.* 32 : 49-58.

Bonga, J. M and Durzan, J. (1982) Tissue culture in forestry. Martinus Nijhoff, The Hague, Boston,London.

Brand, M. H. and Lineberger, R. D. (1992) In vitro rejuvenation of *Betula* (Betulaceae): biochemical evaluation. *Am. J. Bot.* 79 : 626-635.

Capuli, F. A. and Calinawan N. M. (1999) Micropropagation of *Eucalyptus* species: research success in a private laboratory. *Canopy Int.* 10-11.

Crawford, D. F and Hartney V. J. (1987) Micropropagation of *Acacia mangium* and *Acacia stenophylla*. In : *Proc ICIAR Intl. Work. on Australian Acacias in Developing Countries.* Australian Centre for International Agriculture Research, Gympie, Australia, pp. 64-65.

Darus, H. A. (1991) Micropropagation of *Acacia mangium* from aseptically germinated seedlings. *J. Trop. For. Sci.* 3 : 204-208.

Darus, H. A. (1992) Micropropagation techniques for *Acacia mangium x Acacia auriculiformis*. In : *Proc Intl. Workshop Breeding Technologies for Tropical Acacias*. ACIAR Malaysia, pp. 119-121.

Fernandez-Lorenzo, Rigueiro J. L. and Ballester, A. (1999) Polyphenols as potential markers to differentiate juvenile and mature chesnut shoot cultures. *Tree Physiol.* 19 : 461-466.

Huang, H. J., Chen, Y., Kuo, J. L., Tzeng, C. C., Huang, B. L., Chen, C. M. and Huang L. C. (1996) Rejuvenation of *Sequoia sempervirens in vitro*: changes in isoesterases and isoperoxidases. *Plant Cell Physiol.* 37 : 77-80.

Huang, L. C., Weng, J. H., Wang, C. H., Kuo, C. I and Shieh, Y. J (2003) Photosynthetic potentials of in vitro-grown juvenile, adult, and rejuvenated *Sequoia sempervirens* (D. Don). *Bot. Bull. Acad. Sin.* 44 : 31-35.

Huang, L. C., Kuo, C. I., Wang, C. H. Murashige T. and Huang T. C. (2000) Ethylene evolution by juvenile and adult developmental phases of *Sequoia sempervirens* shoots cultured in vitro. *Bot. Bull. Acad. Sin.* 41: 263-266.

Ide, Y., Watanabe, Y. and Ikeda, H. (1994) Tissue culture of *Acacia auriculiformis* using the aseptically germinated seedlings. *J. Japanese. For. Soc.* 76: 576-583.

Kaur, K and Kant, U. (2000) Clonal propagation of *Acacia catechu* Willd. by shoot tip culture. *Plant Growth Regul.* 31 : 143-145.

Kuo, J. L., Huang, H. J., Cheng, C. M., Chen, L. J., Huang, B. L., Huang, L. C. and Kuo T.T. (1995) Rejuvenation in vitro: modulation of protein phosphorylation in *Sequoia sempervirens. J. Plant Physiol.* 146: 333-336.

Marguerite, Q., da Silva Mônica, C., Martins Kelly, G. and de Oliveira Dulce, E. (2001) Multiplication of juvenile black wattle by microcuttings. *Plant Cell Tissue Organ Cult.* 66 : 199-205.

Mokotedi, M. E. O. (1999) Studies on the in vitro rooting ability of cold-tolerant hybrids of *Eucalyptus grandis* x *nitens* and assessment of root quality. In: *Forest Biotechnology'99 Conference.* University of Oxford, Oxford.

Murashige, T. and Skoog, F. (1962) A revised medium for rapid growth and bioassay with tobacco tissue culture. *Physio. Plant.* 15: 473-497.

Rey, M., Tiburcio, A. F., Díaz Sala, C. and Rodríguez, R (1994) Endogenous polyamine concentrations in juvenile, adult and in vitro reinvigorated hazel. *Tree Physiol.* 14 : 191-200.

von Aderkas, P and Bonga. J. M. (2000) Influencing micropropagation and somatic embryogenesis by manipulation of phase change and culture environment. *Tree Physiol.* 20 : 921-928.

Watanabe, Y., Ide, Y. and Ikeda, H. (1994) Plant regeneration from axillary bud culture of one-year-old seedling of *Acacia auriculiformis* grown in green house. *Bull. Tokyo Univ. For.* 92:29-35.

❏❏❏

Forest Biotechnology in India © 2008
Editors: S.A. Ansari, C. Narayanan & A.K. Mandal, pp.237-250
Satish Serial Publishing House, Delhi

23

Clonal Micropropagation of Subtropical Fruit Trees

MANEESH MISHRA, RAJESH PATI AND RAMESH CHANDRA

Biotechnology laboratory, Central Institute for Subtropical Horticulture, Rehmankhera, P.O. Kakori, Lucknow-227 107, India

ABSTRACT

Aonla (*Emblica officinalis* Gaertn.), bael (*Aegle marmelos* Corr.) and guava (*Psidium guajava* L.) are three very important fruit crops owing to their neutraceutical properties, wider adaptability under wasteland conditions and increasing demand in pharmaceutical (*aonla* and *bael*) as well as food processing industry. *Aonla* and *bael* are propagated through softwood grafting and budding whereas guava is generally propagated by air-layering, cutting and grafting. Though many new elite cultivars of these fruit crops have been developed, they could not reach end users because of lack of proper propagation system. An effective and rapid *in vitro* regeneration system, using shoot bud culture from mature 'plus trees' of these fruit crops, have been developed at Central Institute for Subtropical Horticulture, Lucknow. The micropropagated plants, which have been acclimatized and transplanted, are performing well in the field.

Key words: Aonla, *Aegle marmelos*, bael, *Emblica officinalis*, guava, micropropagation *Psidium guajava*.

INTRODUCTION

Aonla or Indian gooseberry (*Emblica officinalis* Gaertn.), belonging to the family Euphorbiaceae, is well known for its medicinal and therapeutic properties. The demand of Indian gooseberry fruit is on the rise in domestic as well as in international market due to its well known curative properties (Pathak, 2003). The fruit is one of the richest sources of vitamin C, and its juice contains nearly 20 times more vitamin C than orange juice. The fruits are also rich in minerals, calcium, phosphorus, nicotinic acid, riboflavin, tryptophan, methionine, lysine and possess herbicidal properties (Anonymous, 1960). Besides, several alkaloids, flavinoids, gueinons, steroids and terpenoids have been isolated from Indian gooseberry. Tannins like gallic acid, ellagic acid present in fruit retards the oxidation of vitamin C. On the other hand, tree bark, fruits and leaves are also rich in tannins and polyphenolic compounds. It thrives well in saline sodic soils and also adapted to other wasteland situations (Ram, 1982). *Bael* (*Aegle marmelos* Corr.) belongs to the family Rutaceae and is an important medicinal fruit tree. The demand of this fruit crop is increasing due to its curative properties. The fruit pulp contains marmelosin, which is a laxative, diuretic, and a strong dose of this compound is used as cardiac depressant. Bael is being used in nearly 60 patented drugs in India. The fruits and roots of bael possess antiamoebic and hypoglycemic activity (Chopra *et al.*, 1956; Ponnachan et al., 1993). The alkaloid 'ageline' present in the leaf is a potent antiasthmatic agent (Harvey, 1968). *Bael* tree can be grown under various wasteland situations. Guava (*Psidium guajava* L.), belonging to family Myrtaceae, is native to tropical America, but cultivated in every tropical and subtropical country of the world. In India, guava is well adapted to almost all the states. The major guava-growing states are Bihar, Uttar Pradesh, Karnataka, Gujarat and Andhra Pradesh. It is estimated that the area and production of guava in India are 1,50,900 ha and 1710.6 M ton, respectively (Singh *et al.*, 2003). Many workers have reported protocols for micropropagation of guava (Amin, 1986; Jaiswal and Amin, 1987; Amin and Jaiswal, 1988; Papadatou *et al.*, 1990; Prakash and Tiwari, 1993, 1996) using shoot buds. However, reports on acclimatization, field establishment and *ex vitro* growth of micropropagated plants are not available. Availability of certified planting material is the main limiting factor in the establishment of the above trees. *Bael* and guava are being propagated through soft wood grafting. However, non-availability of desired number of shoots for grafting or budding is another constraint in field multiplication, which is labour intensive and economically not viable due to slow rate of multiplication and low field survivability. The use of micropropagation approach for accelerating the production of clonal stock of commercial cultivars in *aonla, bael* and guava is an important area of research which needs attention. We report here clonal micropropagation protocols for these fruit trees.

MATERIALS AND METHODS

The present investigation was carried out at Biotechnology Laboratory, Division of Crop Improvement & Production, Central Institute for Subtropical Horticulture (Indian Council of Agriculture Research), Lucknow, India.

Aonla

A 14-year-old tree of *aonla* (cv. NA-6) was pruned to stimulate fresh shoots. The excised shoots from the tree brought to the lab were defoliated and trimmed to 3 cm, retaining at least one axillary bud from 10-15th nodal segment of shoot as per protocol of Mishra *et.al.* (1998). The shoots were first washed under running tap water and treated with a solution containing 0.1 % Carbendazime (Bavistin), 60.75 mM Rifampicin and two drops of Tween-20 for 1 h. The treated shoots were further surface sterilized with 0.1-0.3 % $HgCl_2$ and 10-30% NaOCl for different durations followed by 5-6 washings with sterile distilled water.

Indeterminate shoots (2 to 3 cm long) were inoculated on MS medium supplemented with different levels of GA_3 (0 , 1.44, 2.88 and 4.33 mM) and constant regime of kinetin (13.9 mM) and glutamine (342.11 mM) to examine the proliferative ability of shoot buds. Prior to autoclaving the pH of the media was adjusted to 5.75. Microshoots that developed were further multiplied by subculturing them on fresh medium every 4 weeks. Data on various morphogenetic characters (length and number of indeterminate and determinate shoots, number of leaves per determinate shoot) were recorded after four weeks.

In vitro raised microshoots (2-3 cm long) were further cultured on MS and ½ MS basal medium containing IBA (0, 39.36 and 49.20 mM) and NAA (0, 5.37, 10.74 and 16.11 mM) in various combinations with 0.1% activated charcoal. The cultures were kept in dark for first 48 h and finally shifted to light. After 45-50 days, the rooted plantlets were washed thoroughly with sterile distilled water and transferred to glass bottles (450 ml) containing autoclaved coconut husk, sand, soil and FYM (1:1:1) or only sand and soil (1:1). All treatments were supplemented with ½ MS nutrient salt solution and 3.40 mM paclobuterazol.

Bael

A 15-year-old elite tree was pruned to encourage growth of new shoots. Young shoots were excised from the tree, defoliated, and cut into 3 cm length explants each having one axillary bud. The shoots were first washed with tap water and then treated with a solution containing 0.1 % Carbendazime (Bavistin), 25 mg/l Rifampicin and two drops of Tween-20, for one hour. The explants were then surface sterilized with 0.1 % $HgCl_2$ for 6 minutes followed by 5- 6 washing with sterile distilled water. Subsequently, the explants were inoculated on MS medium (Murashige and Skoog, 1962) fortified with two different cytokinins (BAP and kinetin) with varying concentrations (0, 1, 2 and 3 mg/l) for *in vitro* shoot bud induction. The micro shoots thus produced were transferred to MS medium supplemented with varying levels of cytokinins (0, 1, 2 and 3 mg/l) to examine their proliferating ability of microshoots. For *in vitro* rhizogenesis, the shoots were inoculated on semi-solid MS medium (0.8 % agar) supplemented with IBA (5, 10, 15, 20 or 25 mg/l) and sucrose (20 g/l). Before autoclaving the pH was adjusted to 5.7. The cultures were incubated under 3000 lux light illumination (16 h cycle) at 25±2°C temperature and 55 % RH. There were 25 replicates for each treatment and each experiment was repeated thrice. The observations were recorded at weekly intervals and data subjected to statistical analysis using completely randomized design.

Guava

A 3-year-old tree of guava (cv. Allahabad Safeda), raised inside shaded net house and regularly sprayed with fungicide Carbendazim (0.5 %), was used for the study. Shoots were excised from the above experimental tree, brought to the laboratory, defoliated, and cut into 3-4 cm length shoots. The shoots were initially washed with tap water, and then with a solution containing 0.1 % Carbendazim (Bavistin), 25 mg/l Rifampicin and two drop of Tween-20, for one hour. After thorough washing with sterile water the explants were surface sterilized with 0.1 % $HgCl_2$ for 6 minutes followed by 5-6 washing with sterile distilled water. The processed explants were inoculated on MS medium (Murashige and Skoog, 1962) fortified with two different cytokinins (BAP and kinetin) with varying concentrations (0, 1, 2 and 3 mg/l) for *in vitro* shoot bud induction. The medium was solidified with 0.8 % agar and supplemented with 20 g/l sucrose. The pH was adjusted to 5.7 before autoclaving. The cultures were incubated under

4000 lux light illumination (16 hours cycle) at 25±2°C and 55 % RH. There were 25 replicates for each treatment and each experiment was repeated thrice. The micro shoots obtained were transferred to MS medium supplemented with varying levels of cytokinins, viz., BAP and kinetin (0, 1, 2 and 3 mg/l) to examine the proliferating ability of microshoots. The established shoots were then transferred to half and full strength MS media fortified with different concentration of IBA (5, 10, 15, 20 and 25 mg/l) for *in vitro* rhizogenesis. The rooted plants were transferred to glass bottles (450 ml volume) containing autoclaved soil, sand, FYM, coconut husk and vermiculite. All the treatments were supplemented with half strength MS nutrient solution and 1.0 mg/l paclobutrazol.

Hardening and field planting

After initial incubation inside culture room for 15 days, the rooted plants were shifted to a plant growth chamber where light intensity was 10,000 lux and humidity 70-80 %. After 25-30 days, the plants were shifted to shaded net house (50 % light) with misting. The plants were then shifted to pots. The data on acclimatization was recorded after 45 days. Once the plants were hardened, they were transplanted to the field during the month of July- August.

Statistical analysis

To test significance of different treatments, the data were subjected to a two way factorial analysis with completely randomized design.

RESULTS AND DISCUSSION

Aonla

In vitro browning and in-borne contamination were two major problems in establishing aseptic cultures of *aonla*. Our earlier investigation revealed that sealing the cut end of shoot explant with paraffin wax significantly reduced *in vitro* oxidative browning of media, but delayed *in vitro* bud induction and subsequent proliferation (Mishra *et al.*, 1998). Selection of young shoots excised from new flush, pre-treatment with 100 mg/l PVP (Polyvinylpyrollidone) followed by their frequent sub-culturing were found to be effective in controlling problem of browning. Observations revealed that maximum survival (66.49%) and minimum infection (33.51%) were achieved when explants were treated with 0.1% $HgCl_2$ for 8 minutes. Higher concentrations of $HgCl_2$ (0.2 and 0.3 %) were detrimental to explants.

Our observations are in accordance with the findings of David *et al.* (1978), Laxmi Sita *et al.* (1979), Lakshmi Sita and Shoba Rani (1985). A perusal of data (Table 1) clearly revealed that GA_3 was critical for proliferation of shoots. MS medium supplemented with 4.33 mM GA_3, 13.9 mM kinetin and 342.11 mM glutamine supported the highest shoot proliferation (13.33 indeterminate shoots/ culture) followed by treatment with 2.88 mM GA_3, 13.9 mM kinetin and 342.11 mM glutamine (10.33 indeterminate shoots/ culture). Moderately hard nodal segments gave the highest *in vitro* bud induction and profuse proliferation than very hard nodal segments (data not shown). *In vitro* bud induction was achieved as per protocol suggested by Mishra *et al.* (1998). Indian gooseberry tree is characterized by phyllanthoid branching habit with two types of shoots. On the basis of growth characteristics, these have been characterized as long (indeterminate) and short (determinate) shoots (Bajpai, 1965; Ram, 1971). Soon after *in vitro* bud induction, a determinate shoot with leaves, flowers appears. However, indeterminate shoots are mandatory for further multiplication. Therefore, role of GA_3 for elongation of indeterminate shoot is of paramount importance. High frequency of shoot proliferation (13.33 shoots/ culture) was observed under the influence of 4.33 mM GA_3, 13.9 mM kinetin and 342.11 mM glutamine. In some woody species, GA_3 has been used in shoot proliferation medium to improve shoot elongation (Wachok and Slius 1980). Usually, GA_3 inhibits shoot bud formation. However, once shoot buds were formed it did not inhibit their further development (Thorpe and Meir, 1973). *In vitro* leaf fall was the major problem in micropropagating aonla. Incorporation of glutamine (342.11 mM) solved this problem. Fortification of amino acids and polyamines in the medium to check *in vitro* leaf fall has already been reported in many crops. Amino acids such as glutamine, alanine, arginine and proline have been found to play important role in prevention of senescence of leaf (Galston, 1983; Montagwe *et al.*, 1978; Bradley *et al.*, 1984).

Microshoots were cultured in MS and ½ MS supplemented with different combinations of IBA and NAA and 0.1% activated charcoal. Results clearly indicated that microshoots cultured on MS medium supplemented with 39.364 mM IBA and 5.37 mM NAA rooted profusely (9.67 roots per explant) (Table 2). However, microshoots inoculated in ½ MS containing 49.2 mM IBA and 210.74 mM NAA produced the longest roots (3.04 cm) having fibrous root architecture without any callus phase, whereas MS supplemented with 39.364

Table 1: Effect of various phytohormones on *in vitro* shoot proliferation in aonla (*Emblica officinalis*).

Treatment (mM)	Length of shoots (cm)		No. of shoots		No. of leaves
	Indeter-minate	Deter-minate	Indeter-minate	Deter-minate	
MS	0.800	0.000	6.000	0.000	2.333
MS+GA_3 1.44 +Kinetin 13.9 + Glutamin 2.88	0.967	0.067	7.667	0.667	3.667
MS+$GA_3$2.88 +Kinetin 13.9+Glutamin 2.88	1.167	0.167	10.333	1.000	4.333
MS+$GA_3$4.33 +Kinetin 13.9+Glutamin 2.88	1.633	0.400	13.333	3.000	5.667
SE ±	0.10	0.05	0.89	0.46	0.51
CD (0.05)	0.231	0.115	2.055	1.062	1.21

Table 2: Influence of auxins on *in vitro* rhizogenesis in aonla (*Emblica officinalis*)

Bioregulants (mM)	No. of roots	Length of roots (cm)	Root wt. (gm)	Root diameter (mm)	Days taken for rooting
MS+IBA 39.36+ NAA 5.37	9.67	1.94	0.0160	0.0200	6.33
MS+IBA 39.36 + NAA 10.74	3.67	2.45	0.0200	0.0167	7.00
MS+IBA 39.36 +NAA 16.11	6.30	2.17	0.0051	0.0267	7.00
MS+IBA 49.20 +NAA 5.37	3.67	2.96	0.0170	0.0367	5.67
MS+IBA 49.2 +NAA 10.74	3.67	2.58	0.1880	0.0233	6.67
MS+IBA 49.2 +NAA 16.11	3.67	1.99	0.0183	0.0233	7.00
1/2MS+IBA 39.36 +NAA 5.37	3.33	2.83	0.0046	0.1330	7.67
1/2MS+IBA 39.36 +NAA 10.74	4.33	1.82	0.0798	0.0167	8.67
1/2MS+IBA 39.36 +NAA 11.16	3.67	2.10	0.0117	0.0467	8.67
1/2MS+IBA 49.20 +NAA 5.37	1.00	2.73	0.0058	0.0233	7.33
1/2MS+IBA 10.0 +NAA 10.74	4.00	3.04	0.0098	0.0333	7.33
1/2MS+IBA 10.0 +NAA 16.11	1.00	1.33	0.0020	0.0200	8.67
SEm±	0.88	0.51	-	0.0038	-
CD (0.05)	1.82	1.056	NS	0.0080	NS

mM IBA and 5.37 mM NAA produced the thickest roots (root dia = 0.1330 mm) lacking root hairs. All microshoots cultured on MS medium, irrespective of hormonal regime employed, produced small amount of callus at the shoot base, which affected the quality of feeder root. Such plants did not acclimatize well. It can be concluded that ½ MS (49.20 mM IBA + 210.74 mM NAA) proved to be better for *in vitro* rooting of Indian gooseberry. No callus growth was observed at the base of callus. Many sub-tropical fruit crops have been found to produce *in vitro* roots under the influence on IBA and NAA combination. Roy *et al.* (1993) reported effectiveness of ½ MS salt medium (supplemented with 5.37 μM NAA and 4.92 μM IBA) for *in vitro* rooting in jackfruit (*Artocarpus hetrophyllus*). The use of these two auxins for better rooting was suggested by earlier workers (Amin and Jaiswal, 1987; Kornova, 1995). The cultures were kept in the dark condition for first 48 hours and then exposed to light for normal growth of roots.

Maximum mortality of micropropagated plants occurs during acclimatization phase because plantlets undergo a rapid and extreme change in physiological functioning when they are removed from *in vitro* condition to *in vivo* condition. In order to devise an efficient acclimatization procedure, it was imperative to investigate some of the carriers or substrates that can influence survival of micropropagated plants. Out of three different substrates used, autoclaved soil, sand and FYM (1:1:1) fortified with ½ MS salt solution and 3.40 mM paclobutrazol was found ideal for acclimatization of *aonla* (survival rate = 72.22%) compared to other substrate like coconut husk. Paclobutrazol, a growth retardant is known to facilitate acclimatization of many fruit crops. Micropropagated plantlets of grown on coconut husk showed premature leaf fall whereas they grew well on a mixture of soil, sand and compost. Rahman (1988) found that *Artocarpus hetrophyllus* plantlets survived much better if they are grown for the first 20 days in compost containing either low levels of nutrients or no nutrients.

Bael

Among the cytokinins (BAP and kinetin) tested, *in vitro* bud induction was first observed in MS medium supplemented with 3.0 mg/l kinetin followed by BAP. All the variables tested differed significantly. As many as 9 microshoots per explant were recorded under the influence of kinetin (3 mg/l). Interestingly, BAP was found to be associated with higher shoot proliferation of micro shoots (Table 3). BAP at 3 mg/l concentration produced 4.6 shoots per explant. From one explant, 3088 plants were regenerated in four-sub culture cycle. Repeated and quick subculturing produced healthy shoots devoid of callus. Comparatively higher rates of multiplication of shoots using microshoots and stem explants on MS medium containing BAP, kinetin and NAA has been reported by several investigators (Hossain *et al.*, 1993; Arya *et al.*, 1981). The reason for such high *in vitro* regenerabilty of bael showed in earlier reports (Arya *et al.*, 1981; Hossain *et al.*, 1993; Aurumugam and Rao, 1996; Islam *et al.*, 1996 a, b) might be due to the juvenile nature of explant. Efficient *in vitro* rooting is a crucial factor for successful regeneration of any plant. *In vitro* rhizogenesis was achieved using higher concentration of IBA and microshoots rooted easily on MS medium supplemented with 25 mg/l IBA (Table 4).

Table 3 : Effect of cytokinins on *in vitro* shoot proliferation in *bael* (*Aegle marmelos*).

Cytokinins (mg/l)	Shoot/explant	Leaves	Days taken for Bud burst
BAP (0.0)	1.33	3.0	8.0
BAP (1.0)	4.0	16.0	7.0
BAP (2.0)	5.0	21.0	6.1
BAP (3.0)	6.33	29.0	5.0
Kin (0.0)	1.33	3.0	8.0
Kin (1.0)	6.0	15.0	5.0
Kin (2.0)	7.0	19.33	6.0
Kin (3.0)	9.33	24.0	5.0

Table 4 : Influence of IBA on *in vitro* rhizogenesis in *bael* (*Aegle marmelos*).

IBA (mg/l)	No. of roots	Root length (cm)	Root weight (mg)	Root diameter (mm)	Days taken for root initiation
10	0.0	0.0	0.0	0.0	
15	1.33	2.63	102.0	0.416	30-45
20	1.66	3.80	191.6	0.573	30-45
25	2.33	3.83	198.3	0.570	25-30
SE ±	0.288	0.146	2.51	0.274	
CD (0.05)	0.940	0.476	8.18	0.895	

The rooted plants which were maintained on coconut husk supplemented with ½ MS nutrient solution and paclobutrazol (1.0 mg/l) under mist condition, showed maximum survival (83.3 %). The growth characters of plants significantly improved when grown on coconut husk as compared to other carrier substrates. This may be due to enhanced porosity as well as better water holding capacity of coconut husk.

Guava

Data revealed that three cm long shoot explant inoculated on MS medium fortified with 3.0 mg/l BAP exhibited quicker bud induction (5 days) and better *in vitro* shoot proliferation (2.67 shoots/culture) (Table 5). Frequent subculturing was needed for removal of phenols. *In vitro* shoot proliferative efficiency of guava was observed to be very low (2.67 microshoots/culture). Amin and Jaiswal (1988) also reported that nodal segments taken from *in vitro* proliferated shoots of guava gave rise to 2-4 shoots by precocious axillary branching without an initial lag-period. By repeated sub-culture, a large number of shoots were produced with a shoot multiplication rate of 3 to 4 fold per sub-culture. Papadatou *et al.*(1990) reported establishment of proliferating shoot cultures from shoot tips (excised from seedlings grown in a growth chamber) on Rugini Olive medium (OM) supplemented with benzyladenine (BA). In the present study, maximum proliferation was achieved with 2 mg/l BA. The low multiplication rate may be due to low BA concentrations (1 mg/l). *In vitro* rhizogenesis (Table 6) was achieved when microshoots were transferred to MS medium fortified with IBA. Maximum roots (3 roots per shoot) could be induced in approximately 17 days with 10 mg/l IBA. *In vitro* rooting was observed to be better in full strength MS than half strength. However, some workers have reported better rooting in half strength MS (Amin and

Table 5: Effect of cytokinins on *in vitro* shoot proliferation in guava (*Psidium guajava*).

Cytokinins (mg/l)	Days taken for shoot bud induction	Shoot length (cm)	Number of shoots	Number of leaves	Culture weight (gm)
0.0 BAP	15.67	0.17	1.0	0.0	0.0023
1.0 BAP	8.33	0.57	1.0	4.67	0.0208
2.0 BAP	6.67	0.73	1.0	5.67	0.0224
3.0 BAP	5.00	0.67	2.67	5.0	0.0208
1.0 Kinetin	4.67	0.57	2.0	4.0	0.0204
2.0 Kinetin	6.33	0.27	1.0	2.0	0.0046
3.0 Kinetin	7.33	0.33	1.0	2.67	0.0057
SE±	0.98	0.23	0.50	0.75	0.0019
CD (0.05)	1.74	0.41	0.87	1.32	0.0034

Jaiswal, 1988). Maximum survival of micropropagated plants (84 %) was noticed on autoclaved coconut husk fortified with ½ MS nutrient solution containing 1 mg/l paclobutrazol. However, successful acclimatization of rooted shoots have also been reported in peat-based compost and autoclaved FYM and sand (1:1) by various workers (Amin and Jaiswal, 1988; Prakash, 1992; Prakash and Tiwari, 1993, 1996). Hardened plantlets were shifted in field where they are growing vigorously.

Table 6: Effect of strength of medium and IBA on *in vitro* rhizogenesis in guava (*Psidium guajava*)

IBA (mg/l)	Days taken for rooting	Number of root	Length of roots (cm)	Weight of roots (gm)
Half MS + 2.5 IBA	31.0	1.33	1.967	0.0047
Half MS + 5.0 IBA	34.33	1.33	3.017	0.0067
Half MS + 10.0 IBA	29.67	1.33	2.283	0.0051
Half MS + 25.0 IBA	29.33	2.00	2.960	0.0058
Full MS + 2.5 IBA	22.33	2.33	2.950	0.0089
Full MS + 5.0 IBA	23.33	2.33	1.890	0.0077
Full MS + 10.0 IBA	17.67	3.0	6.750	0.0211
Full MS + 25.0 IBA	30.67	2.33	1.523	0.0080
Full MS+ 0.0 IBA	0.00	0.00	0.000	0.0000
SE±	0.220	0.36	0.0	2.0
CD (0.05)	0.45	0.74	0.0	4.12

References

Amin, M.N. (1986) Micropropagation of a tropical fruit tree *Psidium guajava* L. *J. Indian Bot. Soc.* 65: 43-44.

Amin, M. N. and Jaiswal, V.S. (1987) Rapid clonal propagation of guava through *in vitro* shoot proliferation on nodal explants of mature tree. *Plant Cell Tissue Organ Cult.* 9 : 235-243.

Amin, M. N. and Jaiswal, V. S. (1988) Micropropagation as an aid to rapid cloning of guava cultivars. *Sci. Hort.* 36: 89-95.

Anon. (1960) Raw Material. In: *Wealth of India - Vol. 13-C.* CSIR, New Delhi, India.

Arumugam, S. and Rao, M. V. (1996) *In vitro* production of plantlets from cotyledonary node cultures of *Aegle marmelos* (L) Corr. *Adv. Plant Sci.* 9: 181-186.

Arya, H. C., Ramawat, K. G. and Suthar, K. C. (1981) Culture and differentiation of plants of economic importance of *Aegle marmelos* L. *J. Indian Bot. Soc.* 60: 134 -137.

Bajpai, P. N. (1965) Studies on vegetative growth on development of male and female gametophyte in aonla (*Emblica officinalis* Gaertn.). *Agra Univ. J. Res. Sci.* 14 : 167- 86.

Bradley, P. M., Fiki, F. and Gites K. L. (1984) Polyamines and arginine after somatic embryogenesis in *Daucus carota*. *Plant Sci. Lett.* 34:397-401.

Chopra, R. N., Nayar, S. L. and Chopra, J.C. (1956) *Indian medicinal plants, 2nd edition*. Bishen Singh Mahendra Pal Singh, Dehradun.

David, H., Semukali, K. and Dadid, A. (1978) Obtention de plants de pin martime (*Pinus pinaster*) a portir de branchy blast and' apex canlinaires de trees jeunes sujets cultivar *in vitro*. *Acad. Sci. Press* 287 : 245-248.

Galston, A. W. (1983) Polyamine as a modular of plant development. *Biol. Sci.* 33 : 382-388.

Harvey, S. K. (1968) A preliminary communication of action of *Aegle marmelos* (bael) on heart. *Indian J. Med. Res.* 56 : 327.

Hossain, M., Karim, R., Islam, R. and Joarder, O. I. (1993) Plant regeneration from nucellar tissue of *Aegle marmelos* through organogenesis. *Plant Cell Tissue Organ Cult.* 34 : 199.

Islam, R., Ahad, A., Rahman, M.H., Hossain, M. and Joarder, O. I. (1996a). High frequency adventitious plant regeneration from radicle explants of *Aegle marmelos* Corr. *Pak. J. Bot.* 28 : 2203- 2206.

Islam, R., Hossain, M., Reza, M. A., Mamun, A. N. K. and Joarder, O. I. (1996b) *J. Hort. Sci.* 71: 995-1000.

Jaiswal, V. S. and Amin, M. N. (1986) *In vitro* shoot proliferation and plant formation from somatic tissues of guava. In: *VI Intl. Congress on Plant Tissue Culture*. University of Minnesota, Minneapolis, p. 279

Kornova, K. (1995) Study of *in vitro* rooting in some peach varieties "Rasteniev". *Dni-Nauki*. 32 (7-8) : 109-111.

Lakshmi Sita, G. and Shobha Rani, B. (1985) *In vitro* propagation of *Eucalyptus grandis* L. by tissue culture. *Plant Cell Rep.* 4: 63-65.

Laxmi Sita, G., Ram, N. V. R. and Vaidyanath, C. S. (1979) Differentiation of embryos and plantlets from shoot callus of sandal wood. *Plant Sci. Lett.* 15 : 265-270.

Mishra, M., Tiwari, R. and Chandra, R. (2003) Micropropagational studies in bael (*Aegle marmelos Corr)*. In: 2nd *Int. Conference on Plant Physiology*. IARI, New Delhi, pp. 512.

Mishra, M., Saxena, R. P., Pathak, R. K. and Srivastava, A. K. (1998) Effect of antioxidants, phenol binding agents and explant waxing on *in vitro* browning of aonla (*Emblica officinalis*). *Prog. Hort.* 30 (3-4) : 128-139.

Montagwe, M. L., Kopperbrink, J. W. and Jaworski, E. G. (1978) Polyamines metabolism in embryonic cells of *Daucus carota*.1.Changes in intracellular content and site of synthesis. *Plant Physiol.* 62 : 430-433.

Murashige, T. and Skoog, F. (1962) A revised medium for rapid growth and bioassay with tobacco tissue culture. *Physiol. Plant.* 15 : 473-497.

Papadatou, P., Pontikis, C. A., Ephtimiadox, E. and Uydaki, M. (1990) Rapid multiplication of guava seedlings by *in vitro* shoot tip culture. *Sci. Hort.* 45 : 99-103.

Pathak R. K. (2003) *Status report on genetic resources of Indian gooseberry-aonla (Emblica officinalis Gaertn) in South and South East Asia.* IPGRI Office for South East Asia, New Delhi.

Ponnachan, P. T. C., Paulose, C. S. and Panikkar, K. R. (1993) Effect of leaf extract of *Aegle marmelos* in diabetic rats. *Indian J. Exp. Biol.* 31: 345-347.

Prakash, H. and Tiwari, J. P. (1993) Micropropagation of Guava (*Psidium guajava*). In: *Proc. of an Intl. Conf.on Biotechnology in Agriculture and Forestry.* New Delhi.

Prakash, H.and Tiwari, J. P. (1996) *In vitro* clonal propagation of guava (*Psidium guajava* L.) cv. Sardar. In: *National Symposium on Horticulture Biotech.* Indian Institute of Horticultural Research, Bangalore, pp. 4.

Prakash, H. (1992) *Micropropagation of Guava (Psidium guajava).* Ph.D. Thesis, G. B. Pant University of Agriculture and Technology, Pant Nagar, India.

Rahman M. A. (1988) Effects of nutrients on the growth and survival of *in vitro Artocarpus heterophyllus* Lam. plantlets after transfer to *ex vitro* conditions in the glasshouse. *J. Hort. Sci.* 63 : 329-336.

Ram S. (1971) *Studies on physiology of fruit growth in aonla* (*Emblica officinalis* Gartn). Ph.D.Thesis, Kanpur University, Kanpur, India.

Ram S. (1982) *Aonla (E.officinalis Gaertin) - Uses, Botany and culture.* Directorate of Experiment Station. G. B. Pant University of Agriculture and Technology, Pant Nagar, India .

Roy, S. K., Islam, M. S., Sen J., Hossain, A. B. M. E. and Hadiuzzaman S. (1993) Propagation of flood tolerant jackfruit(*Artocarpus hetrophyllus*) by in *vitro culture. Acta Hort.* 336 : 273-278.

Singh, G., Rajan, S. and Singh, A. K. (2003) Approaches and strategies for precision farming in guava. In: *Precision farming in Guava* (eds Singh H.P., Singh G.,

Samuel, J. and Pathak, R.K.) NCPAH, DAC/PFDC, CISH, Lucknow, pp 92-113.

Thorpe T. A. and Meir D. D. (1973) Effect of gibbrellic acid and absicic acid on shoot formation in tobacco callus culture. *Physiol. Plant.* 29 : 121-124.

Wachok R. J. and Slius C. J. (1980) *In vitro* propagated and establishment of waxes current (*Ribes inebrians*). *J. Hort. Sci.* 55 : 355-357.

❑❑❑

Forest Biotechnology in India © 2008
Editors: S.A. Ansari, C. Narayanan & A.K. Mandal, pp.251-261
Satish Serial Publishing House, Delhi

24

Standardization of Micropropagation Technique for *Ailanthus excelsa* Roxb. Adult Tree

N. K. SHARMA, R.K. GUPTA
AND UTTAR K. TOMAR

Forest Genetics & Tree Breeding Division, Arid Forest Research Institute
New Pali Road, P.O. Krishi Mandi, Jodhpur-342 005, Rajasthan, India

ABSTRACT

Different experiments were carried out to standardize *in vitro* culture establishment, shoot multiplication, rooting and hardening stages of micropropagation in *A. excelsa*. The *in vitro* shoot cultures from coppice shoots on MS medium were established and multiplied for more than two years on MS with 2 mgl^{-1} BA. Average multiplication rate was four fold after every subculture. 50% rooting was achieved on half strength MS medium containing 1.5 mgl^{-1} NAA. Rooted plants were successfully transferred to soil after hardening process.

Keywords : *Ailanthus excelsa*, clonal propagation micropropagation, tree improvement.

INTRODUCTION

Ailanthus excelsa Roxb., a valuable tree species belonging to family Simaroubaceae, is naturally distributed in India, Myanmar, Burma, Pakistan, Bangladesh, and Malaysia (Roxburgh, 1795). It is one of the most important tree species of arid and semiarid areas of India, suitable for sand stabilization, shelterbelts and field boundaries in agroforestry (Anonymous, 1990). It grows well in regions where rainfall ranges from 400-1900 mm (Seth *et al.*, 1962). It can grow on a variety of soils, except clay soil with poor drainage. It grows well on porous sandy loam. The wood of this species is used as door frames, match boxes and packing cases. The leaves are highly palatable and nutritious fodder for sheep and goats. The leaf protein obtained from *A. excelsa* was found superior to soyabean protein (Singh *et al.*, 1992). The seed oil is considered to have insecticidal properties. Bark of this tree is used for the treatment of bronchitis, cold, cough, skin diseases, diarrihea, and dysentery. The crude drug preparation from fresh juicy bark has been proved to be an effective contraceptive (Atique *et al.*, 1993).

The tree is mainly dioecious with some monoecious individuals (Roxburgh, 1795; Dirr, 1990). Flowers are usually unisexual, but bisexual flowers do occur in some individuals (Feret, 1973). Therefore, some trees do not bear seeds, resulting in fodder production different from the types of trees described above. High diversity in terms of growth and fodder value has also been observed in seedling-raised plantations. Hence, fodder production can be increased through cloning of selected superior germplasm. Unfortunately, this tree is very difficult to propagate clonally by conventional methods (Sharma and Tomar, 2003; Tomar, 2003). Present paper deals with the standardization of micropropagation technique for mature trees of *A. excelsa*.

MATERIALS AND METHODS

Nodal segments were collected from branches of 7- to 10-year-old trees, new sprouts from lopped trees, and coppice shoots. These explants were washed with tap water mixed with 2-3 drops of commercial detergent Tween-80 to remove dust particles. This was followed by 6-7 times washing with distilled water to remove the detergent. Thereafter, explants were surface sterilized with 0.1% $HgCl_2$ (w/v) for 3 min in a laminar airflow cabinet, washed 3-4 times with autoclaved distilled water to remove traces of $HgCl_2$ and cultured on MS basal medium (Murashige and Skoog, 1962) or MS medium same containing 1.0 mgl^{-1} BA. All media contained 3% sucrose as a carbon source and were gelled with 0.8% agar (Hi-Media). The pH of the medium was adjusted to 5.8 before autoclaving. The cultures were incubated under 16 h photoperiod at 26 ± 2°C. After establishment of aseptic shoot cultures, further experiments for shoot proliferation were conducted with shoots derived from coppice segments. Different concentrations of cytokinins and auxins were added individually or in various combinations. An experiment was also conducted with single stem node (0.5 cm long) explants derived from *in vitro* proliferating shoots. These single nodal explants were placed in two different orientations (horizontal and vertical position) on shoot multiplication medium containing 0.5 to 8.0 mgl^{-1} BA. For rooting experiments, various concentrations (0.25, 0.5, 1.0 and 1.5 mgl^{-1}) IBA and NAA with ½ strength MS medium were tried. Rooted plantlets were transferred into small plastic cups or glass bottles containing soilrite as a potting mixture and kept in growth chamber under 90% relative humidity, 28 ±2°C temperature and 16 hour photoperiod for initial one week. The relative humidity was decreased gradually every week and finally, in the forth week, it was maintained at 60% level. Half-strength MS liquid medium was provided to rooted plantlets for initial 15 day of growth. Observations were recorded on shoot initiation, shoot multiplication, rooting percentage and root length after 20 days of inoculation.

RESULTS

Physiological status of the explants plays an important role in bud break under *in vitro* conditions. Different explants, viz., nodal segments from branches of mature trees, new sprouts from lopped trees, and coppice shoots were collected and studied for *in vitro* responses. Main constraints faced with the explants collected from mature trees were: 1) thickness of stem branches, which is the main hurdle in handling the explants, 2) secretion of sticky mucilaginous substance from cut ends, which some times causes heavy infection, and 3) stem hairs. Heavy infection rate was observed in cultures from some of the trees. During rainy season infection rate was very high. Finally, some trees which showed less infection were selected. Few trees were lopped and some were cut from the base (15 cm from the ground) to induce coppice shoots.

Establishment of shoots culture

First experiment was conducted with nodal segments derived from branches of normal tree, new sprouts from lopped tree and coppice shoots. All type of stem nodes were cultured on MS media with and without 1mgl^{-1} BA. Nodal segments from branches of mature tree produced maximum exudation of mucilaginous substances, followed by new sprouts from lopped trees. Coppice shoots exudated minimum secondary products into the medium. Only explants from coppice shoot produced shoots (Figure 1 A, B) and other explants failed to produce shoots (Table 1). It was found that explants type played an important role in bud break at initial stage. However, number of shoots increased on BA supplemented media and over all growth of shoots were also better.

Table 1: Morphological response of explant collected from lateral branches of mature tree (MT), lopped tree (LT) and coppice shoots (CS), cultured on MS medium. (Experiment was repeated two times).

Medium	Source and explant number	Thickness of nodal segments (cm)	Mucilaginous substance secretion	Bud break (%)	Remarks
MS* MS* + BA	MT40 MT40	2.0-2.5 2.0-2.5	+++ +++	0 0	No callusing; browning of explant
MS* MS* + BA	LT40 LT40	2.0-2.5 2.0-2.5	+++ +++	0 0	No callusing; browning of explant
MS* MS* + BA	CS40 CS40	0.5-1.0 0.5-1.0	+ +	100 100	Callusing from cut ends

* Basal medium

Shoot proliferation

Shoots derived from coppice shoots were sub-cultured on MS basal medium with 1mgl^{-1} BA and 1mgl^{-1} kinetin to select better combination for shoot proliferation. This preliminary experiment indicated that cytokinin supported shoot growth while BA supported shoot proliferation as well as shoot elongation. Shoots raised through coppice segments were maintained as shoot stock on MS medium containing 1 mgl^{-1} BA. Initially multiplication rate was very poor but after few subcultures (4-5) they showed almost four-fold multiplication rate (Fig.1 C). Shoot cultures were multiplied and maintained on MS with 1 mgl^{-1} BA for further rooting experiments carried out after two years. Shoot multiplication rate was maintained during this long-term subculture period (Figure 1 D). Addition of IAA (0.1–1.0 mgl^{-1}) in MS with 1 mgl^{-1} BA did not influence shoot multiplication rate but shoot length was significantly influenced by addition of 0.25 mgl^{-1} IAA. On the other hand, NAA (0.1–1.0 mgl^{-1}) in MS with 1 mgl^{-1} BA did not influence shoot elongation significantly. On the contrary, NAA significantly increased shoot multiplication rate at 0.25 mgl^{-1} concentration.

Fig. 1 : A & B, Shoot induction from nodal segments excised from coppice shoots of *A.excelsa* on hormone free MS medium; C, Shoot proliferation after few successive subcultures on MS with 1 mgl^{-1} BA; **D,** Multiple shoot which were maintained and multiplied (four fold times) for more than two years during long-term subculture period.

Effect of explants orientation and BA concentration

Nodal shoot segments (0.5 cm) were prepared from multiplied shoot stock. Such segments were placed in horizontal (H) and vertical

(V) positions on MS media containing various concentrations (0.5 – 8 mgl^{-1}) of BA. Data were recorded after 45 days of inoculation and two-factor analysis of variance was performed to investigate the effects of different BA concentrations and two types of orientation for the shoot multiplication and shoot length separately. Shoot multiplication significantly varied under different BA concentrations. However, interaction of both the factors was not found significant. Therefore, only means for main effects (BA concentrations* and orientations**) were compared. Maximum shoot multiplication was recorded in medium with 2.0 mgl^{-1} BA (12.4 shoots) and it was at par in medium with 4.0 mgl^{-1} BA (10.50) and in medium with 8.0 mgl^{-1} BA (10.32). Vertical orientation (11.0) was significantly superior than horizontal orientation (6.47). In the absence of BA, no significant difference was recorded in shoot multiplication and shoot elongation in any of the orientation.

Maximum (1.60 cm) shoot length was observed in vertically oriented shoot buds on MS with 2.0 mgl^{-1} BA (Table 2). Shoot length

Table 2: The effect of orientation of explants and different concentrations of BA on *in vitro* shoot growth of *Ailanthus excelsa* after 45 days of inoculation, when shoot buds cultured on MS medium.

BA (mgl^{-1})	No. of explants	Orientation*	Mean no. of shoots per explant ± SE	Mean shoot length ± SE (cm)**
0.00	16	H	1.25 ± 0.14	0.52^{ab} ± 0.05
	12	V	3.08 ± 0.67	0.55^{ab} ± 0.07
0.50	27	H	3.93 ± 0.69	0.87^{bc} ± 0.12
	18	V	10.53 ± 1.65	1.40^{d} ± 0.13
1.00	21	H	5.80 ± 0.48	1.25^{cd} ± 0.20
	10	V	11.62 ± 2.63	1.23^{cd} ± 0.18
2.00	17	H	8.92 ± 1.72	0.65^{ab} ± 0.06
	13	V	15.84 ± 3.19	1.60^{d} ± 0.31
4.00	18	H	8.00 ± 1.56	0.50^{ab} ± 0.08
	14	V	14.36 ± 2.82	0.41^{ab} ± 0.06
8.00	12	H	10.90 ± 1.27	0.39^{a} ±0.05
	14	V	9.83 ± 1.98	0.42^{ab} ± 0.04

* H- Horizontal; V- Vertical

** Means having common superscripts in the column are not significantly different at the 5% level as separated by Duncan Multiple Range Test (DMRT).

was almost similar in vertically oriented shoot buds in MS medium containing 0.5 mgl^{-1} BA (1.40 cm.), horizontally oriented shoot buds in MS with 1.0 mgl^{-1} BA (1.25 cm) and in vertically oriented on MS with 1.0 mgl^{-1} BA (1.23 cm.). Overall analysis showed that vertical placement of explant on the medium was the best for getting maximum shoot multiplication and elongation.

In vitro rooting

Experiments indicated that at a constant concentration of NAA (1 mgl^{-1}), ½ MS medium was more effective than full strength MS medium. So ½ MS was selected as basal medium for further rooting experiments. IBA and NAA, in various concentrations (0.25, 0.5, 1.0 and 1.5 mgl^{-1}) in ½ strength MS medium, were tried for rooting. Rooting was not achieved in ½ MS without anxin and ½ MS with IBA (0.25 - 1.5 mgl^{-1}). Lower concentration (0.25 mgl^{-1}) of NAA also did not help in rooting. However, poor (16.7 %) rooting was recorded on 0.5 - 1.0 mgl^{-1} NAA (Table 3). Highest (50%) rooting along with callusing was observed on ½ MS medium containing 1.5 mgl^{-1} of NAA (Figure 2 A, B).

Table 3: Effect of auxins on rhizogenesis of shoots of *Ailanthus excelsa* when cultured on ½ strength MS medium supplemented with different auxins. Data were recorded after 45 days of inoculation.

Auxins (mgl^{-1})	No. of explants	Rooting %	Mean no. of roots ± SE	Mean root length ± SE (cm)
0.00	20	0.0	0.0	0.0
IBA (0.25)	10	0.0	0.0	0.0
IBA (0.50)	10	0.0	0.0	0.0
IBA (1.50)	10	0.0	0.0	0.0
NAA (0.25)	10	0.0	0.0	0.0
NAA (0.50)	40	16.71	2.92 ± 0.30	3.96 ± 0.29
NAA (1.50)	30	50.00	1.29 ± 0.16	2.76 ± 0.16

Hardening

Over all success in hardening was only 10% mainly due to frequent failure of power supply to growth chamber (Fig.2 C). Hardened plants were healthy (Fig. 2D).

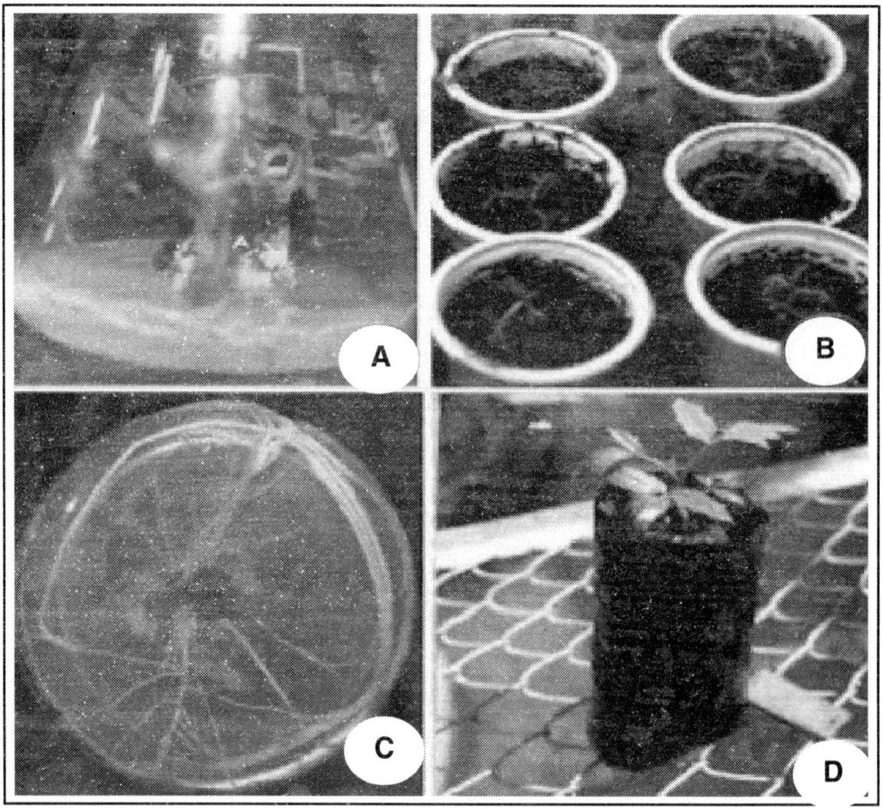

Fig. 2 : A & B, Root induction and growth from isolated micro shoots on ½ strength MS with 1.5 mg/l NAA after 45 days of inoculation; **C,** Hardening of *in vitro* raised plantlets in growth chambers for one month; **D,** One-month-old plant of *A. excelsa* transferred to soil mixture and kept in shade house

DISCUSSION

Choice of explant is a key factor for establishing morphogenic culture in woody species (Vasil and Vasil, 1986; Bonga, 1987). Generally, micropropagation is easy through juvenile explants. Regenerate capacity decreases with age of donor tree. In most of woody species establishment of shoot cultures through explants derived from mature trees is difficult. Mature tree produce juvenile coppice shoots. In the present study, nodal segments collected from mature trees exuded phenolic substances in the medium which inhibit bud break. In contrast, success was achieved in establishing the shoot culture from the coppice shoots. Similar observation have also been made in neem, *Eucalyptus* and many other tree species.

Shoots initiated from coppice shoot explants were very small and took time (three to four subcultures) for attaining good growth in terms of shoot multiplication and shoot length. Improved performance after several subcultures was observed in cultures of *Betula platyphylla* (Mc Cown, 1989). In the course of our investigation, the shoot cultures were maintained for more than two years, without any degradation in multiplying capacity. BA proved superior to kinetin for multiplication and elongation, because naturally occurring cytokinins, are readily oxidized by the cytokinin oxidases, resulting in poor morphogenesis, whereas the synthetic ones remain unaffected (Mok *et al.*, 1982). Similarly, BA proved more effective than kinetin in *Ailanthus malabarica* (D'silva and D'souza, 1992), *Albizia lebbeck* (Upadhyay and Chandra, 1983; Rao and De, 1987), *A. procera* (Roy and Datta, 1985) and *Leucaena leucocephala* (Dhawan and Bhojwani, 1985).

In most of the woody species rooting of *in vitro* raised shoots poses a formidable problem and it is affected by several interacting factors. Rooting in *Ailanthus* is very difficult. In another member of this genus, *A. malabarica, in vitro* shoots were rooted using two-step methods (D'silva and D'souza, 1992). It is well-established that auxins are the main factors involved in the root formation. Among several auxins, IBA is preferred because it is not easily destroyed by autoclaving or light as is the case of IAA. However, *Ailanthus excelsa* micro-shoots could not be rooted on IBA containing media. In contrast, NAA was found suitable for root induction. Present study optimized a micropropagation protocol for *A. excelsa* by standardizing various stages of *in vitro* shoot culture establishment, shoot multiplication, rooting and hardening. While first two steps, viz., establishment of shoot cultures and shoot multiplication work better, rooting and hardening stages need further investigations for their improvement.

ACKNOWLEDGEMENT

The authors is thankful to Director, AFRI, Jodhpur for providing necessary facilities during the studies. This research work was carried out under the Forestry Research Extension and Education Project for which financial assistance was kindly provided by World Bank. The help rendered by Ms Kusum Parihar in editing photographic plates, is gratefully acknowledged.

References

Anon. (1990) *Plants for Reclamation of Wastelands.* Publication and Information Directorate, CSIR, New Delhi.

Atique, A., Khair, S. and Iqbal, M. (1993) Traditional herbal contraceptives from northwest Uttar Pradesh. In: *Medicinal Plant - New Vistas of Research* (Part1) (eds Govil, J. N., Singh, V. K. and Shamima H.) Today & Tomorrow's Printers and Pub., New Delhi.

Bonga, J. M. (1987) Clonal propagation of mature trees: Problems and possible solutions. In : *Cell and Tissue culture in forestry, Vol. 1 - General Principles and Biotechnology* (eds Bonga, J. M. and Durzan, D. J.) Martinus Nijhf. Pub., Dordrecht.

Dhawan, V. and Bhojwani, S. S. (1985) *In vitro* vegetative propagation of *Leucaena leucocephala* (Lam.) de wit. *Plant Cell Rep.* 4 : 315-318.

Dirr, M. A. (1990) *Manual of woody landscape plants: their identification, ornamental characteristics, culture propagation and uses.* Stipes Publishing Co., 1007 p.

D'Silva, I and D'Souza, L. (1992) Micropropagation of *Ailanthus malabarica* DC using juvenile and mature tree tissues. *Silvae Genet.* 41 : 333-339.

Feret, P. P. (1973) Early flowering in *Ailanthus. For.Sci.* 19 : 237-239

Franclet, A. (1983) Rejuvenation: theory and experiences in clonal silviculture. In: *Symp. On Clonal Forestry, Its Impact on Tree Improvement and our Future Forests,* Toronto, Ontario.

Mc Cown, B. H. (1989) Birch (*Betula* spp.) In : *Biotechnology in Agriculture & Forestry Vol. 5, Trees II* (ed Bajaj, Y.P.S.) Springer Verlag, Berlin.

Mok, M. C., Mok, D. W. S., Dixon, S. C., Armstrong, D. J. and Shaw, G. (1982) Cytokinin structure-activity relationship & metabolism of N6- (?2 - isopentenyl) adenosine - 8-14C in Phaseoulus callus tissues. *Plant Physiol.* 70 : 173-178.

Murashige, T. and Skoog, F. (1962) A revised medium for rapid growth and bioassay with tobacco tissue culture. *Phsiologia Plant.* 15 : 473-497.

Rao, P. V. L. and De, D. N. (1987).Tissue culture propagation of tree legume *Albizia lebbeck* (L) Benth. *Plant Sci.* 51 : 263-267.

Roxburgh, W. (1795) *Plants of Coast of Coromandal - Vol.1.* N. Bulmer and Company, Shakespeare Printing Office, London.

Roy, S. K. and Datta, S. K. (1985) Clonal propagation of a legume tree *Albizia procera* through tissue culture. *Bangladesh J. Bot.* 14 : 127-131.

Seth, S. K., Raizada, M. B. and Khan, W. M. A. (1962) *Trees of Vana Mahotsava.* F. R. I. and Colleges, Dehradun.

Singh, R. P., Gupta, M. K. and Chand, P. (1992) Auto-ecology of *Ailanthus gladulosa* Desf. in western Himalayas. *Indian For.* 118 : 917-921.

Upadhyay, S. and Chandra, N. (1983) Shoot and plantlet formation in organ and callus cultures of *Albizia lebbeck* Benth. *Ann. Bot.* 52 : 421-424.

Vasil, I.K. and Vasil, V. (1986) Regeneration in cereal and other grass species. In: *Cell Culture and Somatic Cell Genetics of Plants, vol. 3* (ed Vasil, I. K.), Academic Press, Inc., New York.

□□□

Forest Biotechnology in India © 2008
Editors: S.A. Ansari, C. Narayanan & A.K. Mandal, pp.263-274
Satish Serial Publishing House, Delhi

25

Economics and Factors Influencing Cost of Micropropagated Plants

U.K. TOMAR, USHA NEGI AND A.K. SINHA

Forest Genetics and Plant Breeding Division, Arid Forest Research Institute, New Pali Road, P.O. Krishi Mandi, Jodhpur 342 005, India

ABSTRACT

Present paper describes economics of large-scale micropropagation of plants, and areas where further research is required to make micropropagation as an economically viable tool in forestry sector.

Key words: biotechnology,cost analysis, tissue culture plants

INTRODUCTION

Micropropagation is a high tech method of vegetative propagation by which selected genotype can be rapidly mass multiplied aseptically under controlled environmental conditions (Murashige, 1974). It is a multi-stage process consisting of atleast four distinct stages: (1) Establishment of aseptic cultures, (2) Shoot multiplication, (3) Rooting of shoots and (4) Hardening (Thorpe and Patel, 1984). In first two stages, stem nodes or shoots are grown on a well defined nutrient medium (MS; frequently used medium) in a culture room or in a growth chamber where light intensity, photoperiod and temperature are controlled (Delbergh and Read, 1991). The primary advantage of micropropagation is the rapid production of quality, disease-free and uniform planting material. The plants can be multiplied continuously through out the year under a controlled environment, irrespective of the season and weather (Murashige, 1974).

However, micropropagation technology is expensive compared to conventional methods and requires several types of skills. It is a capital-intensive industry, and in some cases the unit cost per plant becomes unaffordable. Major portion of the tissue culture plant cost is constituted by labour and electricity. In the developed countries, labour is the major contributor (60 to 80%) towards the cost of tissue-cultured plants (Anderson *et al.*, 1977; Donnan, 1986). Several tissue culture units were established in Asian countries due to availability of sufficient manpower at low costs.

Therefore, it is worth studying cost analysis of tissue culture raised plants in Indian scenario. With this objective, efforts have been made to give an insight into the cost structure of tissue culture plants if produced at commercial scale. The cost calculation is neither feasible nor desirable for a research laboratory where objective is to standardize a protocol. Some equations have also been developed to predict the production of tissue culture plants (TCPs) for a well-defined protocol from a fixed capacity Tissue Culture Plant Production Unit (TCPPU). This paper also describes the methodology of cost estimation, factor analysis and areas where still researches are required to make micropropagation an economically viable tool in forestry sector.

METHODOLOGY

Defining a protocol and plant production schedule

From forestry point of view, a tissue culture protocol of neem (*Azadirachta indca*) has been considered as a model. At AFRI a protocol has been standardized for adult neem trees. Plants have been established in field using the above protocol; neem can be multiplied and maintained in TC lab for more than two years with a consistent shoot multiplication rate of almost four- fold. The various protocol stages are provided in Table 1.

Table 1: Stages of tissue culture protocol, multiplication factor at each stage and time period required for subsequent subculture in neem.

Stage	Multiplication factor	Subculture Time
Initiation of culture	1.0	4x2 weeks
Shoot multiplication	4.0	4 weeks
Rooting	0.8 (80%)	4 weeks
Hardening	0.8 (80%)	4 weeks

In case of trees, when cultures are established, a few subcultures are required to stabilize shoot multiplication rate. Present protocol requires two subcultures to get four-fold shoot multiplication. Therefore, plant production will start after 20 weeks from initiation of the first batch of shoot culture.

Production schedule

If 2000 cultures are initiated every week the production schedule according to above protocol will be in three phases (Table 2) namely, initial middle and end phase. In initial phase, cultures are established, multiplied and rooted to the full capacity of TCPPU. In present case initial phase is 20 weeks and production starts after 20 weeks. In the middle phase we can get a consistent production of plants depending upon protocol efficiency. At the end, when we wish to stop producing plant, all the cultures are transferred to rooting medium, instead of maintaining a stock of shoot multiplication, for continued supply of micro-shoots for rooting. Thus, the production is increased in end phase as compared to middle phase. In Table 2, schedule for production of plants in first year and second year at weekly interval is given for the above-defined protocol.

Table 2 : First year plant production schedule on weekly basis

Wks	Establishment	Shoot Multiplication	Rooting	Hardening	Plants output	Phases
	(1.0)	(4.0)	(0.8)	(0.8)	← Multiplication Factor	
1	2000					
2	2000					
3	2000					
4	2000					
5	2000					
6	2000					
7	2000					
8	2000					
9	2000	2000				
10	2000	2000				Initial phase
11	2000	2000				
12	2000	2000				
13		2000	6000			

Contd...

Table 2. Contd...

14		2000	6000			
15		2000	6000			
16		2000	6000			
17		2000	6000	4800		
18		2000	6000	4800		
19		2000	6000	4800		
20		2000	6000	4800		
21		2000	6000	4800	3840	
22		2000	6000	4800	3840	
23		2000	6000	4800	3840	
24		2000	6000	4800	3840	
Middle phase production schedule will be same till 40th week						
41			8000	4800	3840	Middle phase
42			8000	4800	3840	
43			8000	4800	3840	
44			8000	4800	3840	
45				6400	3840	
46				6400	3840	
47				6400	3840	
48				6400	3840	
49					5120	End Phase
50					5120	
51					5120	
52					5120	
		Total output in a year			128,000	
Two years plant production schedule on weekly basis:						
Middle phase production schedule remains same till 92nd week						
93			8000	4800	3840	Middle phase
94			8000	4800	3840	
95			8000	4800	3840	
96			8000	4800	3840	
97				6400	3840	
98				6400	3840	
99				6400	3840	
100				6400	3840	
101					5120	End Phase
102					5120	
103					5120	
104					5120	
		Total output in a year			327,680	

Defining a model of production unit

As per the above defined protocol and production schedule, we kept cultures for 4 weeks in one stage before transferring them to next stage. Hence, we required 8000 culture in shoot multiplication and 24000 shoots in rooting stage in four weeks time. For the above, a PTCU having a capacity of space for 32000 cultures was needed. A TC laboratory with 12 caster racks accommodated 32000 cultures at a time. Three air conditioners (2 tons) were required to maintain a constant temperature of culture room. Accordingly, we required other facilities and manpower to run the laboratory. Details of the same are given under the heading cost structure (Table 3).

Table 3 : List of equipments and their cost analysis in a year (in Rupees).

Equipments	Quantity	Rate per Unit (Rs.)	Amount (Rs)	Life (Year)	Annual Contribution (Rs.)
Caster rack	12	25000	300000	10	30,000
Laminar flow chamber	3	35,000	105000	10	10,500
pH meter	1	20,000	20000	10	2,000
Electronic Balance	2	25,000	50000	10	5,000
Magnetic stirrer	1	2,000	2000	10	200
Autoclave	1	250,000	250000	10	25,000
Distillation Unit	2	25,000	50000	10	5,000
Minor Equipments	1	58000	58000	10	5,800
Mist Chamber	1	500,000	500000	10	50,000
Shade House	1	100,000	100000	10	10,000
	Total		14,30,000		143500

Cost structure

General structure of cost engrossed cost of equipments, electricity consumption, media (chemicals), glassware, labor and other office expenses. We estimated cost of different constituents required to run a TCPPU having capacity of 32000 cultures with defined protocol as mentioned earlier. In the following paragraphs, estimates are prepared for different categories keeping a concept of TCPPU for producing 128,000 TC plants as per schedule observed for above defined protocol in a year.

Equipments

Necessary equipments required in a TCPPU conceptualized with a capacity of managing 32000 cultures at a time are given in Table 3. All the equipments can be managed for ten years with a maintenance amount of Rs 72,000/- per year (included under office expenses). Therefore, 10 % of total cost of equipments was considered in present case. Cost of building was not included in inputs because property cost does not depreciate.

Electricity

Electricity consumption of above mentioned TCPPU was estimated on the basis of electric equipments / appliances consumption potential in watts, multiplied by numbers, hours (in use) per day and number of day in a year. An estimate is given in Table 4 for electric consumption (about 77,000 units) of a TCPPU per year. Total expenditure on electric bill was computed to be about Rs 3.5 lakhs at the rate of Rs 4.5 per unit (Table 4).

Table 4: Electricity consumption by various appliances used in TCPPU.

	No A	Watts B	Hrs C	Days D	Units (Kwatts)* AxBxCxD/1000
Tissue Culture Equipment					
Tubes lights	144	40	16	365	33638
Air Conditioners	3	3500	8	365	30660
Autoclave and distillation unit	1	9000	2.5	260	5850
Minor equipment	1	200	3	260	156
Green House					
Fans	2	750	4	365	2190
Motors	2	400	4	365	1168
Office and miscellaneous expenses		1000	8	365	2920 76582
			Total Cost		Rs. 2,00,000

* Total units = 76582 and Electricity Cost Rs. 3,44,619 at the rate of Rs. 4.5 per unit;

Chemicals

Total amount of media required is estimated according to the production schedule. One lakh twenty eight thousand (128000) plants are produced if 2 lakhs shoots are transferred in rooting medium at the rate of 64% success (combining rooting and hardening success). Similarly 2 lakh shoots are produced from 64,000 cultures in multiplication stage raised from 16000 initial cultures. This can also be calculated by adding all the figures in shoot multiplication and rooting column in Table 2 (production schedule for one year).

Therefore, total medium required to grow 2,80,000 cultures in a year (16000 initial cultures + 64000 shoot multiplication stage + 2,00,000 in rooting stage) will be around 2800 l considering 10 ml medium required per culture. Thus, 3000 l is considered as the requirement for one year TC plant production (Table 5).

Table 5: Expenditure on chemicals in a year (in Rupees).

Chemicals	Qty (Liter)	Rate	Annual Cost
Media	3000	50	150000
Alcohol	300	100	30000
Misc. Chemicals (PGRs etc)			12000
Insecticide/Fungicide treatment			8000
	Total Cost		Rs. 2,00,000

Glassware

Glassware requirement for about 40000 cultures (maintenance of 32000 cultures for 4 weeks and additional 8000 cultures weekly for transfers) considering one glass bottle with a capacity to grow four cultures (shoots bunch for proliferation or shoots for rooting) with 40 ml of medium for four week period, is estimated and summarized in Table 6. It is assumed (on the basis of our experience) that glassware can be used for at least four years. Some losses in glassware per year can be met from office expenses. Hence, the one-fourth cost of glassware is considered for a year, except polybags.

Table 6: Expenditure on glassware in a year (in Rupees)

Glassware	Quantity	Rate	Amount	Life	Annual Cost
Beakers (10, 5 , 2, 1 liter)	20	200	4000	48	1000
Flat Bottom Flask (10, 5 , 2, 1 liter)	20	2000	40000	48	10000
Petri Plates	50	20	1000	48	250
Measuring cylinder (10, 5 , 2, 1 liter)	20	50	1000	48	250
Measuring cylinder (200,100, 50 ml)	40	25	1000	48	250
Glass Bottle	10000	12	120000	48	30000
Root Trainer 150 CC	1600	120	192000	48	48000
Polythene bags	150000	0.1	15000	12	15000
Miscellaneous					15250
				Total	120,000

Manpower

As per above protocol schedule, 8000 cultures (2000 shoot multiplication + 6000 rooting) are to be transferred (subculured) every week under the defined set up of present TCPPU. If we consider one operator (skilled person) can transfer 500 cultures per day, then 3 operators can transfer 9000 cultures (500 cultures x 3 persons x 6 days) in a week. One skilled labour and one unskilled labour are required for media preparation. Two unskilled persons are required to help operators working on the laminar air flow cabinets for providing old cultures and fresh media at regular intervals. One more person is also required to keep all the records of culture and medium other store items. Two persons are required to maintain hardening areas and nursery. Therefore, a total of 4 skilled and 6 unskilled persons are required to produce TC plants in the above-defined TCPPU. One qualified manager is also required to run TCPPU. Table 7 summarizes the total requirement of manpower and their annual salary.

Table 7: Requirement of manpower and their wages (monthly and annually)

Manpower	No.	Salary/ month (Rs)	Annual salary (Rs.)	Total (Rs.)
I. Manager	1	10,000	1,20,000	5,76,000
II. Skilled	4	5,000	2,40,000	
III. Unskilled	6	3,000	2,16,000	
		Total		5,76,000

Office expenses

A lump-sum amount of Rs 2.0 lakhs (Rupees two lakhs) will be required to run TCPPU office for paying bills of telephone, fax, internet facility, stationary items, equipment and building maintenance and other unforeseen expenses (Table 8).

Table 8 : Subheads under office expenses

Item	Per Month	Per Annum
Office Expenses	5,000	60,000
Equipment Maintenance	6,000	72,000
Miscellaneous (Unforeseen)	5,000	60,000
	Total Rs.	192,000

Cost analysis

Total cost involved in production of 1.28 lakhs plants with a above defined TCPPU and defined protocol is Rs 16.0 lakhs as summarized in Table 9. Major portion of TC plants cost is contributed by manpower (35%) followed by electricity charges (22%). Rest of four categories contributes 43% (about 10% each) (Figure 1; Pie chart).

Table 9: Different categories of in TC plant costs in a year.

	Heads	Rupees (in Lakhs)	Per cent
Major	Man Power	5.80	35
	Electricity	3.50	22
Minor	Chemicals	2.00	13
	Office Expenses	2.00	13
	Equipments	1.50	9
	Glassware	1.20	8
	Total annual cost	**16.00**	**100**

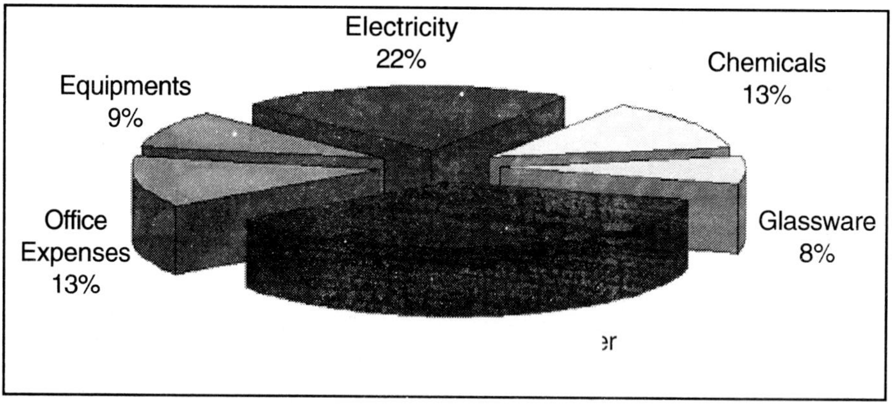

Fig. 1: Pie chart of percentage contribution by different factors (defined categories). For tissue culture production of neem using protocol developed at AFRI, Jodhpur.

Cost per plant

In one year, about 16 lakhs rupees expenditure may incur to produce 1.28 lakhs plants from a above defined TCPPU using this given protocol but in following year production cost will be 1.99 lakhs rupees because second year plant production continues from first week. Therefore, in two years, total production is 3.27 lakhs plants. According to above cost analysis, first year plants will cost Rs 12.5 per plant, but in the second year, it is reduced to Rs 9.8 per plant (Table 10). In other words, cost of non-productive phase (initial phase 20 weeks) is contributing towards the higher cost in first year but there is no non-productive phase in second year, hence, second year cost per plant is less.

Table 10 : Cost per tissue culture plant in first year and second year.

	1st Year	2nd Year
Inputs in Rs. (Lakhs)	16.00	32.00
Production in No. (Lakhs)	1.28	3.27
Cost per plant (Rupees)	12.5	9.8

CONCLUSION

Three main factors influencing the cost of TC plants are protocol efficiency, manpower, and electricity. Other minor factors are chemicals, equipments and glassware. Therefore, there is a great scope of reducing the cost of TC plants if we can minimize cost of major factors. In following paragraphs some suggestion are given to minimize the cost of these three major cost contributors.

Plant tissue culture protocol efficiency influences the cost of TC plants tremendously. Shoot multiplication rate influence the TC plant cost inversely. Higher the multiplication rate, lower is the cost. However, shoot multiplication should be measured on the basis of number of shoots transferred to rooting media per unit time. Less number of stages always reduces the TC plant cost. A TC protocol requires at least two stages, viz., shoot multiplication and rooting cum hardening (*ex vitro* rooting). But, this is very difficult in case of tree species.

Manpower contributes to 70% costs of TC plants, if TC plants are produced in developed countries (Delbergh and Read, 1991). In India, it reduces tremendously and contributes only 35%. But, still manpower contributes a major portion of it. Therefore, establishing a TCPPU in an area where the daily wages rates are cheaper can reduce manpower cost. Better management of manpower will also help in cost reduction of the plants.

Second major factor contributing to high cost is electricity (20%). Lighting of shelves and cooling of culture room consumes major portion (85%) of electricity. Establishing a TCPPU in suitable geographical area of low temperature and high humidity can reduce electricity cost. In such areas low cost poly houses can be useful and electricity consumption by AC will also be reduced. Designing a laboratory utilizing natural sun light can further cut electricity bills. Standard Himedia chemicals costs about Rs. 100 per liter of a medium, whereas media prepared in laboratory will cost less (Rs. 50). In place of costly glassware, plastic wares can be used.

Finally, protocols should be standardized for those species where product value is as high as it can sustain a TCPPU easily and can also earn profit.

ACKNOWLEDGEMENT

The authors wish to thank Director, AFRI for encouraging present studies. The help rendered by Ms Kusum Parihar in typing the manuscript, is gratefully acknowledged.

References

Anderson,W. C. Meagher, G. W. and Nelson, A. G. (1977) Cost of propagating broccoi plants through tissue culture. *Horti . Sci.* 12 : 543544.

Delberg, P. C. and Read, P. E. (1991) Micropropagation. In: *Micropropagation, Technology and Application* (eds Delbergh, P.C. and Zimmerman, R.H.) Kluwer Academic Publisher, Netherland, pp 1-13.

Donnan, A. (1986) Determining and minimizing production Cost. In : *Tissue culture as plant production system for horticulture crops* (eds Zimmerman, R.H. Griesbach, R.J. Lawson, R.H.) Martinus Nijhoff Publ., Dordrecht, pp.167-173.

Murashige, T. (1974) Plant propagation through tissue culture. *Ann. Rev. Plant Physiol.* 25 :135-166.

Thorpe, T. A. and Patel R. K. (1984) Clonal Propagation: Adventitious Buds. In : *Cell Culture and Somatic Cell Genetics Vol. 1* (ed Vasil I. K.) Academic Press, Newyork, pp. 49-60.

□□□

Forest Biotechnology in India © 2008
Editors: S.A. Ansari, C. Narayanan & A.K. Mandal, pp.275-283
Satish Serial Publishing House, Delhi

26

The Role of Fungi in the Production of Aromatic Agarwood in *Aquillaria agallocha* (Roxb.), A Commercially Important Medicinal Tree Species of Arunachal Pradesh

KARUNA SHRIVASTAVA
ANURADHA KUMARI AND TASSO TABIN

Plant-Microbe Interaction Laboratory, Department of Forestry,
North Eastern Regional Institute of Science & Technology
Nirjuli-791109, Arunachal Pradesh, India

ABSTRACT

Aquillaria agallocha (Roxb.), a very precious tree species of North-eastern India, produces resin-impregnated heartwood that is fragrant and highly valuable. Agar is formed as the result of natural infection, found as irregular patches in trunks of the standing tree of *A. agallocha.* Three hypotheses exist regarding agarwood formation, namely through pathological, wounding/pathological and non-pathological processes. A few fungi associated with the wood have been considered to be involved in formation oleoresin deposits in *A. agallocha.* A study was conducted to find out the role of fungi in the production of agar wood in *A. agallocha*. Samples were collected from natural as well as plantations of *A. agallocha* and subjected for the isolation of fungi following standard methodology. 14 different species of fungi were isolated and identified. No wood sample was found free of fungus. More fungal species were recorded in wood of tree from natural forest than from plantations. *Penicillium* spp. dominated followed by *Aspergillus* spp. One species each of *Fusarium, Botryodyplodia, Diplodia, Trichoderma* and *Chrysogenum* was also recorded. *Phialophora parasitica,* which is considered as the main cause of agar wood formation was not isolated from any of the samples.

Keywords : *Aquillaria agallocha,* agarwood, fungi

INTRODUCTION

Aquilaria agallocha (Common name: Agarwood, eaglewood, gaharu; Pharmaceutical name: Lignum Aquilariae) is a medicinally important Indo-Malayan tree distributed in the tropical belt of north-east India, especially in the foothills of Arunachal Pradesh. *A. agallocha* and other species in the genus *Aquilaria* produces agarwood which is resin impregnated, fragrant and highly valuable. The wood is in high demand for making medicine, incense and perfumes in Asia and the Middle East. *Aquilaria* trees are indiscriminately cut in search for resin and agarwood. Eventually, *Aquilaria* species have been threatened and listed in the 2002 IUCN Red Data Book. Unless further actions are taken to control unsustainable agarwood harvest and illegal trade, natural populations of *A. agallocha* and other agarwood-producing species will continue to decline. The negative impacts of such declines will be felt not only in terms of biodiversity loss but also in terms of the reduced availability of valuable forest resource.

Three hypotheses exist regarding agarwood formation, namely result of pathological activities, wounding/injuring and non-pathological processes (Ng *et al.*, 1997). The formation of agarwood may be in response to fungal attack, which requires several years to form (Hawksworth and Gibson, 1977; Jamaluddin, 1977; Gibson, 1977; Rahman and Basak, 1980; Tamuli *et al.*, 2000), wounding (Pojanagaroon and Kaewrak, 1980) or chemical change (Verma, 1977), and the aromatic resin (wood) is usually formed inside the trees with no symptom expression externally. Although research into the origin of agarwood is continuing, it appears that the fragrant resin that permeates the heartwood of some *Aquilaria* trees is produced as a response to wounding and/or a fungal infection. *Aquilaria* trees are naturally infected by a variety of fungi, but the interaction between the host tree and the fungi in order to produce agarwood is poorly understood. Consequently, a specific fungal group responsible for converting simple wood into valuable agarwood has not been identified, which can be used to artificially induce agarwood production in *Aquilaria* trees. Oldfield *et al.* (1998) observed that resin production was due to infection by fungi, viz., *Aspergillus* spp., *Botryodiyplodia* spp., *Diplodia* spp., *Fusarium bulbiferum, F. laterium, F. oxysporum, F. solani, Penicillium* spp. and *Pythium* spp. The major chemical components responsible for the characteristic scent of Agarwood products are sesquiterpenes, which so far could not be produced artificially. Hence production of precious agarwood and agarwood oil depend totally on the natural processes. In India, the best grade agarwood comes from Arunachal Pradesh (Palit, 1996).Characterizing fungal diversity responsible for agarwood formation will be useful for adopting artificial induction program as followed by countries like Vietnam.

Keeping in view the above facts, a study was carried out in order to ascertain whether fungi are associated with the heartwood of *Aquilaria agallocha*.

MATERIALS AND METHODS

Wood samples were collected from different sources (natural forest plantations and local market) in sterilized polythene bags and brought to the laboratory for further investigation. Age of wood samples varied but could not be ascertained with available data. The investigation included microscopic examination and identification of fungal isolates from twenty four samples of agarwood (12 each from natural and plantation). The wood samples were surface sterilized with 0.01% mercuric chloride and inoculated on potato dextrose agar supplemented with chloramphenicol, an antibiotic, to avoid bacterial contamination. Plates were incubated at 25°±2°C for 4 -7 days and observed for fungal growth. The fungal colonies were sub-cultured and identified on the basis of morphological, cytological and microscopic characteristics (Barnett and Hunter, 1972; Ellis and Ellis, 1985, Ellis, 1971, 1976; Subramanian, 1971). To test the pathogenicity, all fungal isolates were used artificial infection of wet agarwood pieces in moist chambers and incubated at 25° ± 2° C, till the initiation of wood infection.

RESULTS AND DISCUSSION

A total of fourteen different fungal species were isolated from twenty four samples of agarwood, out of which eleven species were identified up to species level (Table 1). Two isolates of genus *Penicillium* could not be assigned to their species hence listed as *Penicillium* spp.I and *Penicillium* spp.II respectively. Species of *Diplodia* could also not be identified. All the twenty four wood samples were found infected with fungi (Table 2). A maximum of six different fungal species were isolated from one wood piece (4 %) collected from natural forest (Fig.1). More than half of the wood samples (59%) were found associated with at least two fungal species followed by three and four fungal species in 8% wood samples respectively, and five fungal species in 4 % of the samples. About 17% of wood samples were associated with only one type of fungi (two each collected from natural and plantation sites). Maximum number of fungal species (11) was recorded wood of trees from natural forest compared to plantation (09) of *A. agallocha* (Table 1). The presence of fungi with every agarwood sample indicates that there is probably a relationship between fungi and agarwood formation in *Aquillaria* tree. However, Gibson (1977) investigated four wood samples collected from Sylhet region of Bangladesh, and

Table 1: List of various fungi isolated from agarwood *(A. agallocha)* and their percentage infection in natural forest and plantation.

S. No.	Name of Fungi	Natural forest stand		Plantation	
		Infection	wood samples infected	infection	% wood samples infected
1.	*Aspergillus flavus* Link	+(3)	25.00	+ (5)	41.66
2.	*Aspergillus terreus* Thom	+ (4)	33.33	+ (1)	8.33
3.	*Aspergillus versicolor* (Vuillemin) Tiraboschi	+ (1)	8.33	-	-
4.	*Botryodyplodia* theobromae Patouillard	+ (2)	16.66	-	-
5.	*Chaetoeium globosum* Kunze.	+ (6)	49.98	+ (3)	25.00
6.	*Diplodia* spp.	-	-	+ (2)	16.66
7.	*Fusarium oxysporum* Schlect.	+ (2)	16.66	+ (1)	08.33
8.	*Penicillium chrysogenum* Thom	+ (2)	16.66	+ (4)	33.33
9.	*Penicillium citrinum* Thom	+ (4)	33.33	-	-
10.	*Penicillium nigricans* Bainier	-	-	+ (3)	25.00
11.	*Penicillium stipitatum*	+ (1)	8.33	-	-
12.	*Penicillium spp. I*	+ (5)	41.66	-	-
13.	*Penicillium spp. II*	-	-	+ (1)	8.33
14.	*Trichoderma album* Preuss	+ (4)	33.33	+ (3)	25.00

* + = present; - a = bsent

* The value in parenthesis indicate total number of wood blocks infected.

Table 2: Percentage of wood samples infected with different fungi after artificial inoculation.

S. No	Name of the fungi	Wood infection (%)	Incubation period (in days)
1.	*Aspergillus flavus* Link	33.33	7
2.	*Aspergillus terreus* Thom	20.83	9
3.	*Aspergillus versicolor* (Vuillemin) Tiraboschi	4.16	5
4.	*Botryodyplodia* theobromae Patouillard	8.33	14
5.	*Chaetonium globosum* Kunze.	37.50	20
6.	*Diplodia* spp	08.33	15
7.	*Fusarium oxysporum* Schlect.	12.50	7
8.	*Penicillium chrysogenum* Thom	25.00	6
9.	*Penicillium citrinum* Thom	16.66	6
10.	*Penicillium nigricans* Bainier	12.50	4
11.	*Penicillium stipitatum*	4.16	5
12.	*Penicillium spp. I*	20.83	6
13.	*Penicillium spp. II*	4.16	8
14.	*Trichoderma album* Preuss	29.16	9

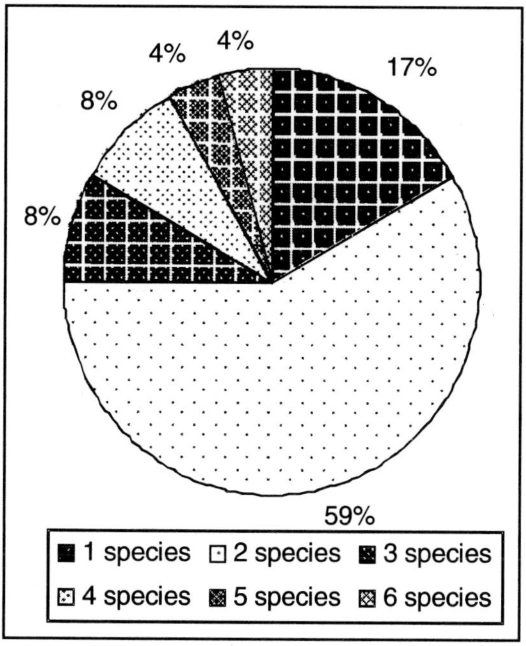

Fig.1: Number of fungal species associated with wood samples.

concluded that it was unlikely that there was a specific fungal cause for agarwood formation. Rahman and Basak (1980) found agar production in trees after artificial inoculation without wounding using three fungal isolates from agar. However, wounding without inoculants, induced colour changes and led to oleoresin production in the host, characteristic of agar. They inferred that formation of agar did not depend on the activity of fungus, but a general reaction of the host to injury or invasion. Verma (1977) did trials with auxin herbicides, viz., 2.4-D, 2.4.5-T and MCPA for inducing formation of agarwood in *A. agollocha* but found that herbicides and the mechanical injury did not appear to induce the formation of agarwood even after one year.

Out of fourteen different species of fungi isolated and identified, *Penicillium* was the dominant genus with six different species followed as *Aspergillus* (three species) and *Fusarium, Botryodyplodia, Diplodia, Trichoderma* and *Chaetonium* one species each (Fig. 2). Jamaluddin (1977) isolated *Cytosphaera manganiferae* Died. from the diseased tissues of standing trees of *A. agallocha.* (Tamuli *et al.*2000) isolated two fungi, viz., *Fusarium oxysporum* Schlect. and *Chaetonium globosum* Kunze. from the diseased wood of *A. agallocha.* When inoculated artificially, these two fungi colonized wood blocks of *A. agallocha.*

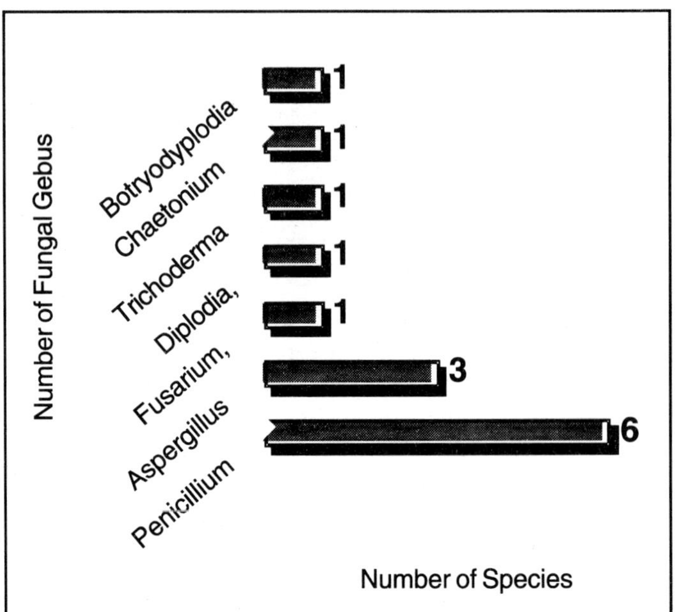

Fig. 2: Number of fungal species recorded.

In the above investigation, *Chaetomium globosum* showed the highest percentage (37.50%) infection of the agarwood pieces (Table 2) followed by *Aspergillus flavus* (33.33%), *Trichoderma album* (29.16%), *Penicillium chrysogenum* (25.00%), *Aspergillus terreus* and *Penicillium spp.I* (20.83% each).Three species, viz., *Aspergillus versicolor, Penicillium stipitatum* and *Penicillium spp. II* exhibited lower percentage infection (4.16%).However, *Phialophora parasitica,* which has been known as the main cause of agarwood formation (Hawksworth and Gibson, 1976) was not isolated from any of the samples.

In natural forest of *A.agallocha, Chaetomium globosum* recorded the highest (49.98%) infection followed by *Penicillium spp.I* (41.66%), *Aspergillus terreus, Penicillium citrinum* and *Trichoderma album* (33.33%).In plantation trees, *Aquilaria agallocha, Aspergillus flavus* was found as the main fungus infecting 41.66% of wood samples followed by *Penicillium chrysogenum* (33.33%) and *Chaetonium globosum, Penicillium nigricans* and *Trichoderma album* (25%) (Table 1).

On artificial inoculation, all the fungi isolated from wood samples infected *A.agallocha* wood. However, incubation period varied with different fungal species (Table 2). *Penicillium nigricans* took maximum time to infect in comparison to *Chaetneium globosum,* which required more time (20 days). This was the fungus that caused the highest percentage of infection of agarwood. In general, *Penicillium* spp. took lesser time for artificial infection of wood than any other genera (Table 2). Tamuli *et al.* (2000) followed Koch's postulates with isolates of *Fusarium oxysporum* and *Chaetonium globosum* on agarwood blocks and found that both were able to infect. The same pattern was observed in the present investigation with these two fungi.

CONCLUSION

There is a positive correlation between presence of fungi and fragrant resinous wood formation in *A. agallocha* trees since no wood block were found free of fungus. No fixed pattern in the number of fungal species/tree or type of fungi/tree was observed. *C. globosum* and *Penicillium* were the most dominant genus with the highest number of species recorded. The topography, soil and micro-climatic conditions of the state are highly favorable for the growth and aromatic resin production in *A. agallocha* tree. It is the need of the hour to develop a method which can initiate agarwood production in *A. agallocha* trees through artificial induction of proper fungal pathotypes or identification of enzyme responsible, so that dependence on oleoresins deposits by other natural processes can be overcome.

References

Barnett, H. L. and Hunter, B. B. (1972) *Illustrated genera of Imperfect fungi.* Burgess Publication Company, Minnesota.

Beniwal, B. S. (1989) Silvical characteristics of Aquilaria agallocha Roxb. *Indian For.* 115: 17-21.

Ellis, M. B. (1971) *Dematiaceous Hyphomycetes.* C. M. I., Kew, Surrey, England, 608 p.

Ellis, M. B. (1976) *More Dematiaceous Hyphomycetes.* CMI, Kew, Surrey, England.

Ellis, M. B. and Ellis, J. P. (1985) *Microfungi on land plants - An Identification Handbook.* Croom Helm Ltd.

Gibson, I.A.S. (1977) The role of fungi in the origin of oleoresin deposits (Agaru) in the wood of *Aquillaria agallocha* (Roxb.). *Bano Bigyan Patrika* 6(1) : 16-26.

Hawksworth, D. L. and Gibson, I. A. S. (1977) *Phialophora parasitica, Desr Pathog Fungi Bact.* 51 : 501-510.

Jalaluddin, M (1977) A useful pathological condition of wood. *Economic Botany* 31 : 222-224.

Ng, L. T., Chang, Y. S. and Azizol, L. K. (1997) A review on agar (gaharu) producing *Aquilaria* species *J. Trop. For. Prod.* 2: 272-285.

Oldfield, S., Lusty, C. and MacKinven, A. (1998) *The Word List of Threatened Trees.* World Conservation Press, Cambridge, U.K.

Palit, M. K. (1996) *Aquliaria agallocha* Roxb. – A commercial agar-oil yielding tree of the north east India. *Arunachal For. News* 14 (4) : 30-33.

Pojanagaroon, S, and Kaewrak (1980) Mechanical Methods to Stimulate Aloeswod Formation in *Aquliaria crassna* (Kritsana) trees. In: *Cultivation, Conservation and Sustainable Use of Aromatic Plants* (eds A. Jatisatienr, T. Paratasilpin, S. Elliott, V., Anusarnsunthorn, D., Wedge, L. E., Craker and Z. E. Gardner) ISHS Leuven, Belgium.

Rahman, M. A. and Basak, A. C. (1980) Agar production in agar trees by artificial inoculation and wounding. *Bano Bigyan Patrika* 9 (1&2) : 97-93.

Subramanian, C. V. (1971) *Hyphomycetes.* ICMR, New Delhi.

Tamuli, P., Boruah, P., Nath, S. C., and Samanta, R. (2000) Fungi from diseased agarwood tree (*Aquilaria agallocha* Roxb.): two new records. *Adv.For. Res,* 22:.182-189.

Verma, V. P. S. (1977) Trials of herbicides for inducing formation of agarwood in *Aquilaria agollocha* Roxb." *Indian Perfumer* 21 (3):151-153.

□□□

Forest Biotechnology in India © 2008
Editors: S.A. Ansari, C. Narayanan & A.K. Mandal, pp.285-290
Satish Serial Publishing House, Delhi

27

Nucleic Acid Based Detection of Associated Pathogen of Witches' Broom of *Dendrocalamus strictus*

V. SURYANARAYANA, PRADEEP KUMAR AND M. KRISHNAREDDY

Department of Forest Protection, College of Forestry
Sirsi 581 401, Karnataka India
Plant Virology unit, Indian Institute of Horticultural Research,
Hessaraghatta, Bangalore, Karnataka, India

ABSTRACT

Dendrocalamus strictus is one of the important species for wasteland land development and posses an high potentiality in meeting raw material requirement of paper and pulp industry. Among various diseases of *D. strictus*, Witches's broom was observed in high proportions in the plantations as well as in natural strands to an extent of nearly 20 per cent. Witches' broom affected *D. strictus* showed symptoms of proliferated axillary shoots as long stiff grassy shoots at the nodal region with clusters of severely reduced narrow leaves appearing like a broom. Since Witches' broom is commonly produced by phytoplasma, mid rib and veins of affected and healthy of *D.strictus* were screened for a phloem bound phytoplasma, using polymerase chain reaction (PCR) technique and witches' broom of *Zizypus jujuba* was used as positive control. Oligonucleotide primer pair [PP1 F2 (5' - AAG GAT TGA TCC TGG CTC SGG AT -3') and PP7-R (5'- CGT CCT TCA TCG GCT CTT-3')] specific to the conserved region of 16S r DNA gene were used to amplify a 537 bp sequence of the phytoplasma. Amplified DNA fragments were obtained after 35 cycles of amplification. The results confirmed the presence of phytoplasma in witches broom affected *D. strictus.*

Keywords : Bamboo, DNA amplification, phytoplasme

INTRODUCTION

Dendrocalamus strictus is one of the important bamboo species gaining importance in wasteland development and in providing raw material for paper and pulp industry. Witches's broom disease of *D. strictus* was observed in high levels proportions in plantations as well as in natural stands.

Phytoplasma detection is normally done through symptomology, light microscopy of stained tissues, fluorescent microscopy, and tetracycline sensitivity test. But the above methods have their own drawbacks. Studies on DNA homology in the highly conserved genes encoding ribosomal RNA and ribosomal protein have shown that phytoplasmas comprise a coherent set distinct from all other prokaryotes (Gunderson *et al.,* 1994). Substantial progress has been made in recent years toward improved detection of phytoplasma associated diseases by the application of polymerase chain reaction (PCR).

The disease is fast spreading possibly through vegetative propagation, insect vectors and tissue culture lines derived from infected tissues. A continuous monitoring and indexing of the phytoplasmal disease is essential for the tree growers and forest nursery managers. Developing a quick and reliable technique for the detection of phytoplasma will help to screen disease free plants for large scale propagation. In view of the above, the present investigation was carried out on PCR based detection of witches' broom phytoplasma.

MATERIALS AND METHODS

Total nucleic acid was extracted using CTAB method following the protocol of Doyle and Doyle (1990). Fresh tissue of midrib and young stem (3 g) was ground into fine powder in liquid nitrogen and 27 ml of CTAB extraction buffer (2% CTAB, 10 mM Tris, 20 mM EDTA, 104 M NaCl; pH 8.0). After incubation at $60^{0}C$ for 10 min, the homogenate was cooled to room temperature, and then extracted with an equal volume of chloroform/isoamyl alcohol (24:1). The mixture was centrifuged at 9,500 rpm for 10 min followed by the collection of aqueous phase using a Pasteur pipette and addition of 0.7 volume of

Isopropanol. Then the mixture was kept at - 20^{0}C for at least 2 h and centrifuged at 13- 15, 000 rpm for 30 min to precipitate nucleic acid. Pellets were then dissolved in TE buffer (pH 8.0) containing RNase at a concentration of 10 µg per ml and incubated at 37^{0}C for 1 h. An equal volume of chloroform/ isoamyl alcohol was added and the mixture was centrifuged at 9,500 rpm for 10 min and aqueous phase eluted out. One volume of Isopropanol and 1/10th volume of 3 M sodium acetate (pH 5.2) was added to precipitate the nucleic acid, which was kept at - 20^{0}C for 2 h. The precipitated DNA was centrifuged at 15, 000 rpm for 30 min and the supernatant was discarded. The air dried nucleic acid pellets were dissolved in 1 ml of 1x TE buffer (pH 8.0) and stored at - 20^{0}C.

DNA amplification

DNA extracts derived from infected and healthy plant tissues as well as positive control DNA from phytoplasma of *Zizypus jujuba,* were analysed by PCR using primer pair PP1 F2 (5′ – AAG GAT TGA TCC TGG CTC SGG AT -3′) and PP7-R (5′- CGT CCT TCA TCG GCT CTT-3′) designed for amplifying DNA fragments from nucleotide position # 1 to # 784 on the phytoplasmal 16S rDNA gene nucleotide sequence. PCR assay was performed with a mixture of 25 µl containing 5 µl of test DNA, 2.5 µl of 10 x reaction buffer, 2.0µl $MgCl_{2,}$ 2.0 µl of dNTPs, 0.5 µl of each primer pair and 0.2 µl of Taq polymerase. DNA samples were amplified in a Gene Amp PCR system 9700 (Applied Biosystems, U.K.). PCR parameters consisted of thirty cycles at 50^0 C for 1 min and extension at 72^0 C for 1.3 min, followed by a final extension step at 72^0 C for 10 min.

The PCR amplification products were electrophoresed in 0.8% agarose gel in 0.5x TBE buffer. The gel was stained with ethidium bromide and bands were compared with the positive control (*Zizypus jujuba* phytoplasma sample) and λDNA Hind III DNA ladder. The gels were documented by Alpha digidoc 1000 system (Alpha Innotech Corporation, USA).

RESULTS

Witches' broom affected *D. strictus* showed symptoms of proliferated axillary shoots as long stiff grassy shoots at the nodal region with clusters of severely reduced narrow leaves appearing like a broom (Fig. 1).

Fig. 1: Healthy (left) and witches' broom affected (right) *Dendrocalamus strictus.*

Using a specific primer set (PP1F2 and PP7R), DNA fragments of about 537 bp were directly amplified from position # 1 to # 1784 on the 16S r DNA gene sequence extracted from infected tissues of witches' broom of *D. strictus*. The band size matched with the positive control from diseased *Z. jujuba*. No amplified DNA bands were observed in samples extracted from healthy plants (Fig. 2).

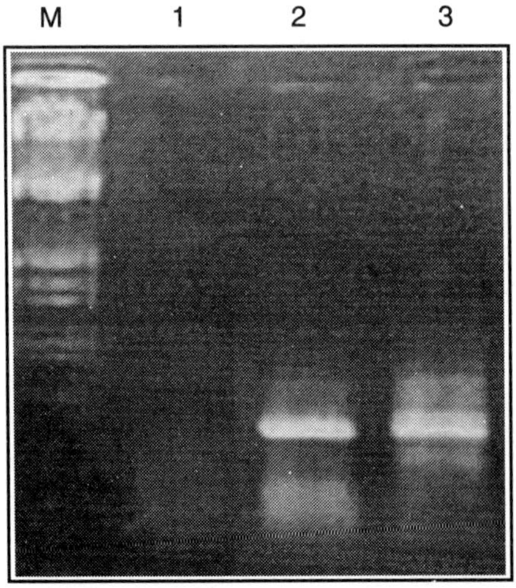

Fig. 2 : Agarose gel electrophoresis of PCR product (16S rDNA Fragment -537 bp) of witches' broom affected *D. strictus*. Lane -M Lambda DNA/ Hind III marker; Lane -1 Healthy; Lane-2 *Dendrocalamus strictus*; Lane-3 Positive control phytoplasma from *Zizuphus jijuba*.

DISCUSSION

The devastating effect of phytoplasma disease of trees has attracted global concern. The present study revealed similar patterns of infection in *D. strictus*. PCR based detection of phytoplasma is preferred since concentration of phytoplasma in hosts is very low. PCR with phytoplasma specific primers have been designed and used for phylogenetic grouping of different phytoplasmas (Lee *et al.*, 1994). In the present study, PCR assay using specific primers was adopted to detect phytoplasma causing witches' broom in *D.* strictus.

In the present study, 16S r DNA was amplified by specific primer pair designed by Lim and Sears (1989). Similar type of molecular detection has been done for phytoplasma of sandal spike (Sunil Thomas and Balasundaran, 1999), *Zizyphus jujuba* (Tian *et al.*, 2000), *Paulownia* (Noriao Shasi *et al.*, 1995), date palm (Husain *et al.*, 2002), lilac (Sinclair *et al.*, 1996) and elm (Jacobs *et al.*, 2003). The present investigation is first of its kind in *D. strictus*.

CONCLUSION

Phytoplasma detection through PCR based techniques can play a vital role in detection and grouping of tree phytoplasmas in India. Owing to its accuracy, rapidity and accessibility, PCR based detection of phytoplasma will also help to screen disease free plants for large scale propagation.

References

Doyle, J. J. and Doyle, J. L. (1990). A rapid DNA isolation. Procedure from small quantities of fresh leaf tissue. *Focus* 12 : 13-15.

Gunderson, D. E., Lee, I. M., Rehner, S. A., Davis, S. A. and Kingsbury, D. T. (1994) Phylogeny of mycoplasma like organisms (Phytoplasmas): A basis for their classification. *J. Bacteriol.* 176 : 5244- 5254.

Husain, A., Al-Awadhi, Asma Hanif, Patrice Suleman. and Magdymontasser, S. (2002) Molecular and Microscopic detection of phytoplasma associated with yellowing disease of date palms *Phoenix dactylifera* L. in Kuwait. *Kuwait J. Sci. Eng.* 29 (2): 87- 105.

Jacbos, K. A., Lee,T. M., Griffith, H. M., Miller, F. D. and Bottner, T. R. (2003) A new member of the Clover proliferation phytoplasma group (16S r VI) associated with elm yellows in Illinois. *Plant Dis.* 87 (3): 241-246.

Lee, I. M., Gundercon, D. E., Hammond, R. W. And Oavis, R. E. (1994) Use of Mycoplasma like organisms (MLOs) group Specific oliigonucelotide primers for nested PCR assays to detect mixed MLO infections in a single host plant. *Phytopathology* 84 : 519-566.

Lim, P. and Sears, B. B. (1989) 16Sr RNA sequences indicates that plant pathogenic Mycoplasma like Organisms are evolutionary distinct from animal Mycoplasma. *J. Bacteriol.* 171: 5906-5911.

Norio Shashi, Hitoshi Nakamura,Nobuyuki Yoshikawa, Takanori Kuono, Trugio Shoji and Tryoshi Takahashi. (1995) Distribution and seasonal variation in detection of Phytoplasma in Bark phloem tissues of single Paulownia trees infected with witches' broom. *Ann. Phytopath. Soc.* 61 : 481- 484.

Sinclair, W. A., Griffiths, H. M. and Davier, R. E. (1996) Ash yellows and lilac witches' broom: Phytoplasma diseases of concern in Forestry and Horticulture. *Plant Dis.* 80 (5) : 468-475

Sunil Thomas and Balasundaran, M. (1999) Detection of sandal spike phytoplasma by polymerase chain reaction. *Curr. Sci.* 76 (12) : 1574 -1576.

Tian, J. B., Guo, H. P., Bertaccini, A., Martini, M., Partrinieri, S. and Partore, M. (2000) Molecular detection of jujube witches' broom phytoplasma in Microporpagated jujube shoots. *Hort. Sci.* 35 (7) : 1274-1275.

Forest Biotechnology in India © 2008
Editors: S.A. Ansari, C. Narayanan & A.K. Mandal, pp.291-298
Satish Serial Publishing House, Delhi

28

Screening Rust Disease Resistance in *Dalbergia sissoo* Through Artificial Inoculation

N. S. K. HARSH, KAMLA UNIYAL AND D. S. BHANDARI

Forest Pathology Division, Forest Research Institute
P.O. New Forest, Dehradun-248006, India

ABSTRACT

Leaf and petiole rust caused by *Maravalia achroa* in *Dalbergia sissoo* is a serious disease, which causes deformity and premature defoliation in young leaves in the early spring (February - March), leading to growth retardation of the plants. For the first time, protocol has been developed for artificial inoculation of rust disease in *Dalbergia sissoo* using urediniospores. The method has been successfully employed for screening 15 clones of *D. sissoo* against rust disease. Clone S-167 (Rajaji National Park Chilla, Kunau range, Uttaranchal) was the most resistant clone, which remained free from infection after inoculation trials. Clones S-57 (Khalawala Range, Ambala Division, Haryana), S-106 (Birdwal range, Hanumangarh Division) and S- 124 (Kosi riverbank, Sunsaria Inerva, Nepal) were also found resistant. Clones S-19 (Shahmansoorpur range, Khanpur Division, Saharanpur, U.P.), S-24 (C. B. Ganj, Bareilly Division, Uttar Pradesh) and S-89 (Hanumangarh range, Comptt. 54 D, Nausand Desal, Shergarh Division, Punjab) were found susceptible.

Key words: Dalbergia sissoo, disease resistance, *Maravalia achroa*, rust,

INTRODUCTION

Due to its valuable timber, *Dalbergia sissoo* Roxb. is extensively planted in avenues, canal, side roadside, and field bunds in private lands, thus playing an important role in meeting the socio-economic needs of the growers. Nearly 10 million saplings are planted every year. The strong and durable heartwood is used for making furniture, construction materials and gun carriage parts. It is naturally distributed throughout the north India growing in the foothills of Jammu and Kashmir, West Bengal as well as in the central and western parts of the Peninsular India.

Leaf and petiole rust caused by *Maravalia achroa* (Syd.) Arthur & Cummins in *D. sissoo* is a serious disease, which causes deformity and premature defoliation of young leaves in the early spring (February – March), leading to growth retardation of the plants. *M. achroa* attacks leaves and juvenile twigs of *D. sissoo* plants. The affected parts are killed resulting in die-back and subsequent death of affected plants. *M. achroa* produces yellowish-orange uredinial sori on lower surface of the leaves and all around the petioles bearing yellow-orange globose to ovate, verrucose uredospores (11 - 15 μm in dia). The brownish telia develop at later stage (September-October) and produce hyaline to light brown teliospores. *Dalbergia latifolia* Roxb. (rosewood) is the other reported host of *M. achroa*.

In recent years, there has been an increasing interest for raising genetically improved planting material for getting timber with high quality. Resistance against diseases and insect pests is one of the criteria for selecting and raising such improved material. Genetically variable seedlings of *D. sissoo* have been produced in Clonal Seed Orchards (CSOs) and Seedling Seed Production Areas (SSPAs) in India. The aim of this study was to develop an artificial inoculation protocol for testing clones of *D. sissoo* seedlings against the rust disease.

MATERIALS AND METHODS

Collection and germination of uredospores

Uredospores of *M. achroa* were collected from infected leaves of field grown *D. sissoo* with the help of a nylon brush and sieved through a 45-micron mesh. The germination of uredospores was tested in sterilized water, fresh leaf extract of *D. sissoo* and 1.5% water agar. The fresh leaf extract was prepared by crushing 5 g young leaves in 25 ml distilled water using a porcelain mortar and pestle and filtering through a Whatman (No.1) filter paper.

The germination trial of uredospores with sterilized distilled water and leaf extract was conducted in glass cavity slides surface sterilized with rectified spirit and in Petri plates (50 mm dia) with water agar. Leaf extract was passed through a syringe filter (Whatman Purodisc™ 25 AS, 0.2 μm pore size) and poured into the cavity slides. Freshly collected uredospores were inoculated in the cavities as well as in the plates with the help of a sterilized nylon brush. The cavity slides were incubated at room temperature (25±2°C) inside a moist chamber prepared with a surface sterilized Petri plate (200 mm dia) with moistened wad of filter papers. The plates were also incubated at the room temperature. Observations on spore germination were taken after every 4 h under the low power of a research microscope as average of five observation fields.

Artificial inoculation of seedlings

The freshly collected and sieved uredospores were used for artificial inoculation studies in two ways. The uredospore inoculum for artificial inoculation studies was prepared either in sterilized water (10 mgl^{-1}) by adding five drops of Tween 80 (0.5%) and soft liquid soap (0.1%) as described by Anderson and Henry (1946), or by mixing equal quantities of uredospores and talc as described by Tervet and Cassell (1951).

Healthy seedlings of *D. sissoo* were selected from the nursery for inoculation by three methods. The leaves and petioles were washed and then sprayed with fine mist of sterilized water through a high-pressure sprayer. In first method, the spore suspension was sprayed on the leaves with an atomizer. In second method, the moistened leaves were wiped to break surface tension on the leaf surface (Dhingra and Sinclair, 1985) and then uredospores-talc mixture applied with the help of a sterile spatula as described by Browder (1971). In the third method, uredospore-talc mixture was applied on the leaves, without wiping them. Three plants were used for each method.

Creation of moist chamber and incubation of inoculated seedlings

A chamber was made with aluminum angle (1.5cm x 1.5 cm) of dimension 90cm x 60 cmx 45 cm and covered with polythene sheet on all sides, except the base. This chamber was set over a sand bed created with sterilized sand moistened with water to create desired high humidity (90±5% R.H.). The inoculated plants were incubated in the chamber for 48 h and then they were kept inside a shade-net house and watered as and when required. Suitable uninoculated controls were also maintained in the same chamber. Observations were taken regularly till symptom development on the leaves and petioles of the inoculated plants.

Screening of clonal material for rest resistance

For comparative assessment of disease development, infected leaves, petioles and rust pustules, were counted. Disease score was worked out as per the formula given below:

Disease score = (No. of infected leaves x Average No. of pustules) + (No. of infected petioles x Av. No. of pustules)

On the basis of disease score, the ranking of clones in different resistant classes was done as mentioned in (Table 1).

Table 1: Scoring of *D. sissoo* clones for disease resistance.

Resistant Class	Disease score
Resistant (R)	0-5
Moderately resistant (MR)	6 -15
Moderately susceptible (MS)	16-25
Susceptible (S)	> 25

Seedlings raised from fifteen clones of *D. sissoo* (Table 2), were inoculated with uredospores and talc mixture in the same manner as described above. Five seedlings of each clone were used as replicates. The experiment was repeated twice.

RESULTS

Uredospore germination

The uredospores started to germinate in leaf extract of *D. sissoo* after 12 h, whereas in water and water agar, the germination started after 24 h. In the leaf extract, germination was complete within 24 hours reaching maximum of 96% germination. In water and water agar, the germination took more than 40 hours with maximum germination was 72%.

Symptom development

Symptoms of rust disease developed only on the seedlings in which inoculation was made with the basidiospore-talc mixture following the second method, whereas no symptoms were developed on the seedlings inoculated by spraying basidiospore suspension and by applying basidiospore-talc mixture without wiping the leaves. Yellowish-orange rust pustules started developing after 10 days of inoculation on the lower surface of leaves and around the petioles, which were similar in form and colour to that of the disease exhibited in the field. The uredospores were observed after 15 days in the pustules, which were similar to those, which were used for inoculation.

Screening of clonal material

Clones S-14 and S-89 were first to exhibit the symptoms of rust disease in the form of pustules whereas clone S-167 remained free from infection. The results of performance of different clones against inoculation of rust are presented in (Table 2).

It is evident from Table 2 that only clone S-167 remained free from rust infection after inoculation with disease score as zero, along with clones S-57, S-106 and S-124 which were also rated resistant with disease score 1. Clones S-19 and S-89 were susceptible of which clone S-19 had the highest disease score of S-186. Clones S-14, S-51 and S-66 were moderately susceptible while the rest were moderately resistant.

DISCUSSION

Germination of uredospores in the three different media established their viability. Determination of viability of spores is an essential requirement before inoculation (Dhingra and Sinclair, 1985). Leaf extract of *D. sissoo* supporting more germination of uredospores indicated its potential as inoculant. This also indicated the receptiveness of the host leaves for infection.

The success of the inoculation method, in developing rust disease in inoculated seedlings, established its effectiveness and further use for the screening of clonal material of *D. sissoo*. Mixing spores in talc and applying them to leaf surfaces is a common technique used in inoculating rust fungi (Dhingra and Sinclair, 1985). The spatula inoculation technique has been reported for inoculation of cereals with rust fungi (Browder, 1971).

Table 2: Performance of different clones of *D. sissoo* against induced rust disease

Clone No./Source	No. of infected leaves	Mean no. of pustules	No. of infected petioles	Mean No. of pustules	Disease score	Resistance class
S-57; Khalawala Range, Ambala Division, Haryana	-	-	1	1	1	R
S-24; C. B. Ganj, Bareilly Division, Uttar Pradesh (U.P.)	5	15	1	1	76	S
S-66; Chhichrauli Range, Yamunanagar Division, Haryana	2	8	1	1	17	MS
S-41; Hasnapur, Tulsipur range, Gonda Division, U.P.	2	4	1	1	9	MR
S-361; Sohelwa Wildlife Division, Gonda, U.P.	2	3	-	-	6	MR
S-10; Pathri Range, Haridwar Division, Uttaranchal	2	5	1	1	11	MR
S-51; Birpur beet, Bhamar range, Gonda Division, U.P.	5	3	1	2	17	MS
S-374; Tulsipur range, Sohelwa Wildlife Division, Gonda, U.P.	3	2	1	1	7	MR
S-106; Birdwal range, Hanumangarh Division	-	-	1	1	1	R
S-124; Kosi riverbank, Sunsaria Inerva, Nepal	-	-	1	1	1	R
S-19; Shahmansoorpur range, Khanpur Division, Saharanpur, U.P.	7	26	1	4	186	S
S-14; Pathri Range, Haridwar Division, Uttaranchal	6	3	2	2	22	MS
S-89; Hanumangarh range, Comptt. 54 D, Nausand Desal, Shergarh Division, Punjab	4	26	1	2	106	S
S-44; Trilokpur, Tulsipur range, Gonda Division, U.P.	1	1	2	6	7	MR
S-167; Rajaji National Park Chilla, Kunau range, Uttaranchal	-	-	-	-	0	R

The development of rust pustules similar to those in field in the present experiment established the effectiveness of the above protocol and experimental conditions. Screening for rust disease in *D. sissoo* through artificial inoculation has not been reported elsewhere. Most of the studies have been done on white blister rust and fusiform rust of pines through artificial inoculations to screen resistant genotypes (Jewell, 1960; Mathews and Rowan, 1972; Sluder and Powers, 1982; Hoff, 1994).

Clone S-167 showing no infection of rust after inoculations emerged to be a resistant clone. Similarly clones with disease score of 1 (S-57, S-106 and S-124), also qualify for resistant category. However, extensive field trials need to be undertaken to confirm the above results.

ACKNOWLEDGEMENTS

The authors are thankful to Dr. S. S. Negi, Director, Forest Research Institute and Dr. A. N. Shukla, Head, Forest Pathology Division, for providing necessary facilities and to Dr. S. Nautiyal, Plant Physiology Discipline, Forest Research Institute, Dehradun for kindly providing the clonal material for this study.

References

Anderson, L. and Henry, B. W. (1946) The use of wetting and adhesive agents to increase the effectiveness of conidial suspensions for plant inoculations. *Phytopathology* 36 : 1056.

Browder, L. E. (1971) *Pathogenic specialization in cereal rust fungi, especially Puccinia recondite f. sp. tritici: Concepts, methods of study and application*. USDA Tech. Bull. No. 1432.

Dhingra, O. D. and Sinclair, J. B. (1985) *Basic Plant Pathology Methods*. CRC Press, Florida.

Hoff, R. (1994) Artificial rust inoculation of white bark pine seedlings- Rust resistance across several populations. *Nutcracker Notes* No. 4.

Jewell, F. F. (1960) Inoculation of slash pine seedlings with *Cronartium fussiforme*. *Phytopathology* 50 : 48-51.

Mathews, F. R. and Rowan, S. J. (1972) An improved method for large scale inoculations of pine and oak with *Cronartium fussiforme*. *Plant Dis. Rep.* 56 : 931- 934.

Sluder, E. R. and Powers, H. R. Jr. (1982) *Fusiform rust infection of loblolly and slash pines after artificial inoculation and natural exposure in plantations.* USDA Forest Service Research Note SE-318, 8 p.

Tervet, I. W. and Cassell, R. C. (1951) The use of cyclone separators in race identification of cereal rusts. *Phytopathology* 41 : 286.

□□□

Forest Biotechnology in India © 2008
Editors: S.A. Ansari, C. Narayanan & A.K. Mandal, pp.299-311
Satish Serial Publishing House, Delhi

29

Biological Control of *Fusarium solani* Through the Interaction of *Trichoderma* and AM Species: An Ecofriendly Approach to Control Vascular Wilt in Neem Seedlings

MUKTA PANDEY, N. S. K. HARSH AND Y. P. SINGH

Forest Pathology Division, Forest Research Institute
P.O. New Forest, Dehradun-248006, India

ABSTRACT

The present study reports the interaction of mycorrhizal and antagonistic fungi for biological control of root diseases in neem (*Azadirachta indica*) seedlings at Forest Pathology Division, Forest Research Institute, Dehradun (Uttaranchal). Application of *Trichoderma* spp. Improved germination, survival, growth and disease reduction either singly or with AM. *Trichoderma koningii* performed better than *Trichoderma harzianum*. The combined application of both biocontrol agents, viz., *Trichoderma* sp. and AM in the presence of pathogen on the host roots was found to be better for biological control as well as growth promotion than their single application. It seems that both of them are compatible with each other in nature. Hence, their combined use is recommended for control of root diseases and promotion of growth in nursery seedlings of *Azadirachta indica.*

Key words: AM fungi, *Azaditrachta indica, Fusarium,* Interaction, *Trichoderma*

INTRODUCTION

Nurseries play a very important role in development of planting stock. Root diseases cause extensive damage to nursery seedlings and result in significant losses in the production of forest produces like timber, forage, fiber, etc. worldwide. Perhaps due to the circumstances that root diseases often remain unidentified and because of the complexity and challenges of working with such diseases, losses in potential production due to root diseases remain largely unquantified.

In many plant diseases the interaction of host, pathogen and physical environment is readily apparent. The recognition of a fourth component, namely other organism(s), which affect the cause of disease either by their direct influence on the pathogen or indirectly through their effect on the host or environment, provides the basis for our concept of biological control. Use of antagonistic microorganisms (organisms that impede the normal activities of the pathogen) is one of the most successful and common methods of biological control. Studies suggests that combined use two or more biological control agents is better. Another type of cross-protection is provided by the symbiosis of the plant and mycorrhizal fungi. Both *Trichoderma* species and Arbuscular Mycorrhizal (AM) fungi, which co-exist in the rhizosphere soils, are known to improve plant growth as well as increase disease resistance against pathogens, but knowledge of *Trichoderma*-AM interactions is inadequate to advise the manipulation of these organisms for optimum plant growth and biocontrol of root diseases. Keeping this in view, the present work was conducted to study the interaction of mycorrhizal and antagonistic fungi for biological control of root diseases in neem (*Azadirachta indica*) seedlings at Forest Pathology Division, Forest Research Institute, Dehradun (Uttaranchal).

MATERIAL AND METHODS

Diseased seedlings of *A. indica* were collected from Central Nursery, Forest Research Institute, Dehradun and a pathogenic isolate P_1 (*Fusarium solani*) was isolated from neem seedlings. Pathogenicity tests according to the Koch's postulates were conducted. Two antagonistic isolates of *Trichoderma* namely, *Trichoderma harzianum* and *T. koningii* were isolated from different substrates (*T. harzianum* from the rhizospheric soil of neem seedlings, Central Nursery, F.R.I., D.Dun, and *T. koningii* from logs of *Pinus radiata* obtained from Wood Seasoning Discipline, F.R.I., Dehradun.) using Soil Dilution Plate Method and identified with the help of standard monographs, keys and descriptions of Bissett (1984, 1991 a, b, c) and Rifai (1969). Mass culture of both selected *Trichoderma* spp. was made in bagasse (Harsh and Ojha, 1998).Mass cultures of *solani* was prepared in CMC medium (Cappellini and Peterson, 1965). Colony forming units (CFUs) of pathogens and antagonists were counted by using a haemocytometer slide.

Glomus fasciculatum was selected for interaction studies with antagonistic fungi. The inoculum of *G. fasciculatum* was multiplied in glass house using trap plants (maize) in pot cultures containing sterilized soil/sand. For assessment of the AM, the soil was subjected to the wet sieving and decanting method (Gerdemann and Nicolson, 1963). The viable AM spores were extracted by sucrose centrifugation technique (Daniel and Skipper, 1982). Staining of roots was done by Phillips and Hayman (1970) method. The degree of mycorrhizal root colonization was calculated by using the gridline intersect method (Giovannetti and Mosse, 1980).

The experiments were conducted at Forest Pathology Division net/ glass house, at Forest Research Institute, Dehradun (Uttaranchal). Duration of net-house experiment for interaction studies was 4 months. In net house experiment, a total of 12 treatments were made in Completely Randomized Block Design (CRBD) in Hiko trays in which AM fungus, antagonistic fungi and pathogen were treated singly as well as in combinations. Uninoculated seedlings served as control. Each treatment had three replications. The soil and sand was collected from Central Nursery, F.R.I. The potting mixture consisted of two parts of sterilized soil, one part sand and 2% farmyard manure. Seedlings were regularly watered with RO filtered sterilized water. The suspension of the pathogen was inoculated one day prior to seed sowing; the biocontrol formulations and AM (*Glomus fasciculatum*) were soil mixed just before seed sowing. Data on percent germination of seeds, mortality of seedlings (disease %) due to damping-off and vascular wilt and survival percentage of seedlings after 4 months were recorded during the experiment

The CFUs of fungal isolates in mass cultures measured by a haemocytometer slide were: *F.solani* = 3.75×10^5CFU/ml, *T. harzianum* (A_6) = 5.94×10^8 spores/ g of bagasse and *T.koningii* (A_7) = 5.38×10^9 spores/g of bagasse.

Treatments made in the glass house experiment were T_1 (Pathogen P_1 alone; P_1), T_2 (Control; C), T_3 (Arbuscular mycorrizia; AM), T_4

(Soil mixing with biocontrol formulation, i.e. A_7, T_5 (P_1 + A_7), T_6 (P_1 + AM), T_7 (A_7 + AM), T_8 (P_1 + A_7 +AM), T_9 (A_6), T_{10} (P_1 + A_6), T_{11} (P_1 + A_6 + AM)] and T_{12} (P_1 + A_6+ AM).

RESULTS AND DISCUSSION

F. solani inhibited the growth of seedlings and was found to be more pathogenic at later stage of seedlings (after 3 months). Diseased plants showed subnormal development because the growth processes were impaired. Mc New (1960) also reported that root diseases by *Fusarium* spp. cause destruction of seedlings via impairment of their growth. *T. harzianum* and *T. koningii* were effective to improve germination and plant growth either singly or in combination with AM even in the presence of pathogen (Table 1) and are in accordance to Windham *et al.* (1986). Arya and Kaushik (2001) also reported improved seed germination due to treatment with *Trichoderma* spp.

The study revealed that some treatments were effective for seed germination when compared with the control (78%) (Table 1). The

Table 1. Effect of different treatments on various parameters of growth of *A. indica*

S. No.	Treatments	Disease %	Germination %	Survival %	Shoot length	Dry shoot wt (g)	Dry root wt (g)
T1	(P1)	76	84	24	8.975	0.153	0.044
T 2	(Control)	0	78		9.65	0.165	0.066
T 3	(AM)	0	70		10.375	0.224	0.072
T 4	(A7)	0	78		12.137	0.3	0.092
T 5	(P1A7)	22	88	74	11.7	0.309	0.095
T 6	(P1AM)	66	84	34	9.437	0.232	0.044
T 7	(A7AM)	0	94		15.237	0.452	0.136
T 8	(P1A7AM)	20	96	80	12.637	0.324	0.12
T 9	(A6)	0	90		10.637	0.244	0.077
T 10	(P1A6)	42	88	58	10.987	0.212	0.044
T 11	(A6AM)	0	80		12.275	0.302	0.095
T 12	(P1A6AM)	32	88	68	11.05	0.205	0.075
	C.D. at 5%	20.906	6.652	20.301	1.534	0.073	0.026

treatments P_1A_7AM (96%), A_7AM (94%), A_6 (90%), P_1A_7 (88%), P_1A_6 (88%), P_1A_6AM (88%), P_1 (84%), P_1AM (84%), and A_6AM (80%) were found to be significantly superior over control, while A_7 (78%) was at par with the control and AM (70%) was significantly inferior to control in terms of germination percentage. However, maximum seed germination occurred in the treatment P_1A_7AM followed by A_7AM and A_6, which were found to be at par with each other. From the point of view of survival percentage, the treatments P_1A_7AM (80%), P_1A_7 (74%), P_1A_6AM (68%), and P_1A_6 (58%) were significantly superior to P_1 (24%), but at par with each other. Results showed that AM was not effective to control disease when applied singly. There is considerable evidence for the role of AM fungi in the control of root pathogens (Caron, 1989; Dehne, 1982; Hooker *et al.*, 1994, St. Arnaud *et al.*, 1995). Disease damage was reduced in 17 out of 32 cases (Dehne, 1982). Some indicated no effect of AM fungi on disease (Baath and Hayman, 1983; Zambolim and Schenck ,1983, 1984). Others reported increased disease severity (Davis *et al.*, 1979; Davis and Menge; 1980). This makes it difficult to draw any general conclusions, partly because so many different pathogens and diseases have been involved, and partly because of the experimental conditions of each study. Clearly, one should expect varied results, even if the AM fungi had been the same (Schenck, 1987).

Both *Trichoderma* spp. significantly reduced percent disease either singly or in combination with AM. There are indications of stimulation of enzyme production by *T. harzianum* to increase the antagonism and thus improving the biocontrol potentiality (Jacobs and Kamoen, 1986). Mayer (1984) observed qualitative as well as quantitative changes in the mycorrhizosphere of endomycorrhizal roots. Hence, modified rhizosphere microorganic population may also interact with pathogens as to inhibit their growth and reproduction. The improved nutrition in mycorrhizal plants leads to decrease in disease incidence (Atilano *et al.*, 1976, Davis and Menge, 1980; Davis *et al.*, 1978). In the present study, modified populations of *Trichoderma* spp. in endomycorrhizal roots seem to influence the pathogen thus resulting in disease reduction.

Kumar (1993) effectively controlled the pre- and post-emergence damping-off of *Pinus roxburghii* in forest nurseries caused by *Fusarium solani* by incorporating *Trichoderma viride* in the potting mixture before sowing the seeds. Harsh *et al.* (1994) successfully controlled the post-emergence damping-off disease of *Moringa pterygosperma* caused by *Fusarium acuminatum* by using *T. harzianum* grown on wheat straw and also by using AM inoculum.

Both antagonists and AM encouraged overall growth of seedlings, when applied singly. Baker *et al.* (1984) found that *Trichoderma harzianum* stimulats growth of plants as a result of production of plant hormones, increased uptake of nutrients by the plant, or the control of one or more subclinical pathogens. Widham *et al.* (1986) also reported plant growth promoting activities due to *Trichoderma* spp. The present results with *Trichoderma* spp. are in conformity to these findings.

It has been reported that mycorrhizal plants develop extensive root system as compared to non-mycorrhizal plants, which ensures the plant with increased availability of water and nutrients, thereby helping the plant for better growth and development (Carling *et al.*, 1978; Jalali and Thareja, 1985; Manjunath and Bagyaraj, 1986; Young *et al.*, 1984). Increased photosynthesis due to mycorrhizal fungi was also reported by Schoenbeck (1979). Davis and Menge (1980) showed that mycorrhizal citrus seedlings infected with *Thielaviopsis basicola* were larger than diseased nonmycorrhizal ones. But the relative reduction of growth of both mycorrhizal and nonmycorrhizal seedlings by *T. basicola* was similar.

Analyses of interaction between pathogen, antagonists and AM

There was a significant difference in dry shoot weight in the presence of AM and antagonist individually as well as in combination of pathogen P_1 x AM x antagonist over control (Table 2).

Table 2: Effect of treatments P_1 (*Fusarium solani*), AM and antagonists individually as well as interactions of their different combinations over control for dry shoot weight of 4-month old neem seedlings.

Source	Type III Sum of Squares	df	Mean Square	F	Sig. at 5%
Corrected Model	0.274	13	2.111E-02	5.527	0.000
Intercept	2.759	1	2.759	722.526	0.000
AM	2.146E-02	1	2.146E-02	5.621	0.027
Pathogen P_1	6.656E-04	1	6.656E-04	0.174	0.680
Antagonist	0.171	2	8.570E-02	22.443	0.000
Replicate	7.030E-03	2	3.515E-03	0.921	0.413

Contd...

Table 2. Contd...

Pathogen P_1* AM	1.193E-02	1	1.193E-02	3.125	0.091
AM * Antagonist	1.837E-02	2	9.184E-03	2.405	0.114
Pathogen * Antagonist	5.314E-03	2	2.657E-03	0.696	0.509
Pathogen P_1* AM * Antagonist	3.821E-02	2	1.911E-02	5.004	0.016
Error	8.401E-02	22	3.818E-03		
Total	3.117	36			
Corrected Total	0.358	35			

In the presence of AM alone, dry shoot weight was increased significantly over the control (Table 3).

Table 3: Effect of mycorrhiza verses control.

Treatment	**Mean Dry shoot wt. in g**
AM	0.301
Control	0.252
SEM ±	0.015
C.D. (at 5%)	0.031

The presence of both antagonists also increased fresh shoot weight significantly over the control. Antagonist A_7 increased more dry shoot weight than A_6 (Table 4).

Table 4: Antagonist v/s control.

Treatment	**Mean (Dry shoot wt. in g)**
Antagonist A_6	0.257
Antagonist A_7	0.369
Control	0.204
SEM ±	0.018
C.D. (at 5%)	0.037

For the interaction of pathogen P_1 x AM x antagonist, the treatment P_1A_7AM was significantly superior to the control while the treatment P_1A_6AM was at par with control for increment in dry shoot weight (Table 5).

It is apparent from the Table 6 that there was a significant difference for dry root weight in the presence of AM and antagonist when they were applied individually.

Table 5: Interaction of pathogen P_1 x AM x antagonists v/s control on shoot weight.

Treatment	AM		Without AM	
	Pathogen P_1	Without pathogen P_1	Pathogen P_1	Without pathogen P_1
Antagonist A_6	0.208	0.298	0.272	0.252
Antagonist A_7	0.336	0.457	0.387	0.297
Without antagonist	0.292	0.217	0.139	0.167
SEM ±	0.036			
C.D. (at 5%)	0.075			

Table 6: Significant difference of the treatments P_1 (*Fusarium solani*), AM and antagonists individually as well as interactions of their different combinations over control for dry root weight of 4-month old neem seedlings.

Source	Type III Sum of Squares	df	Mean Square	F	Sig. at 5%
Corrected Model	4.063E-02	13	3.125E-03	4.658	0.001
Intercept	0.242	1	0.242	360.997	0.000
AM	4.796E-03	1	4.796E-03	7.149	0.014
Pathogen P_1	1.770E-03	1	1.770E-03	2.638	0.119
Antagonist	2.410E-02	2	1.205E-02	17.963	0.000
Replicate	4.674E-03	2	2.337E-03	3.483	0.049
AM * Pathogen P_1	3.352E-05	1	3.352E-05	0.050	0.825
AM * Antagonist	2.217E-04	2	1.109E-04	0.165	0.849
Pathogen P_1 * Antagonist	2.499E-03	2	1.250E-03	1.862	0.179
AM * Pathogen P_1 * Antagonist	2.531E-03	2	1.265E-03	1.886	0.175
Error	1.476E-02	22	6.709E-04		
Total	0.298	36			
Corrected Total	5.539E-02	35			

In the presence of AM alone, dry root weight was increased significantly over the control (Table 7).

Table 7: Mycorrhiza v/s control.

Treatment	**Mean (Dry root wt. in g)**
AM	0.0935
Control	0.0704
SEM ±	0.006
C.D. (at 5%)	0.0124

The presence of both antagonists also increased dry root weight significantly over the control. Antagonist A_7 increased more dry root weight than A_6 (Table 8).

Table 8: Antagonists v/s control.

Treatment	Mean (Dry root wt. in g)
Antagonist A_6	0.0772
Antagonist A_7	0.116
Control	0.0529
SEM ±	0.007
C.D. (at 5%)	0.0145

Table 9 shows that there was a significant difference for percentage root colonization of AM fungus when AM was applied with pathogen P_1 and antagonist individually, but the combination of pathogen P_1 x AM x antagonist had no significant difference over AM.

Table 9: Significant difference of the treatments pathogen P_1 (*Fusarium solani*) and antagonists individually as well as interactions of their different combinations with AM over the treatment AM for per cent mycorrhizal root infection in root cortex of 4-month old neem seedlings.

Source	Type III Sum of Squares	df	Mean Square	F	Sig. at 5%
Corrected Model	1459.096	7	208.442	11.380	0.000
Intercept	76511.027	1	76511.027	4177.146	0.000
Pathogen P_1	150.992	1	150.992	8.243	0.017
Antagonist	1229.534	2	614.767	33.563	0.000
Replicate	38.458	2	19.229	1.050	0.386
Pathogen P_1 * Antagonist	40.112	2	20.056	1.095	0.372
Error	183.166	10	18.317		
Total	78153.288	18			
Corrected Total	1642.262	17			

Increase in percentage root infection of AM was significantly superior for the combination of AM x pathogen P_1 over AM (Table 10).

Table 10: Interaction of pathogen P_1 x AM v/s AM.

Treatment	Mean (% root infection)
P_1AM	68.093
AM	62.300
SEM ±	1.427
C.D. (at 5%)	3.1793

The presence of antagonist A_6 also increased percentage root infection of AM significantly over AM (Table 11).

Table 11: Interaction of antagonists x AM v/s AM

Treatment	Mean (% root infection)
A_6AM	76.365
A_7AM	56.628
AM	62.597
SEM ±	1.747
C.D. (at 5%)	

Results show that colonization of plant roots with AM increases in the presence of *Fusarium solani* and both *Trichoderma* species with respect to the treatment AM alone. Because AM fungi live partly within the host roots, and obtain energy from the plant, they have a competitive advantage over other rhizospheric microorganisms (Davis *et al.*, 1978).Results of this experiment showed the positive impact of *Trichoderma* spp. on germination, growth and disease reduction either singly or with AM and *T. koningii* performed better than *T. harzianum*. In our results, the combined application of both biocontrol agents, viz., *Trichoderma* spp. and AM in the presence of pathogen on the host roots was found to be better for biological control as well as growth promotion than their single application. It also appeared that both of them are compatible with each other in nature. Hence, their combined use is recommended for control of root diseases and the promotion of growth in nursery seedlings.

ACKNOWLEDGEMENTS

We are thankful to Dr. S. S. Negi, Director and Dr. A. N. Shukla, Head of the Division, Forest Research Institute, Dehradun for providing facilities and encouragement during the course of study.

References

Arya, S. and Kaushik, J.C. (2001) Efficacy of *Trichoderma* species & *Gliocladium virens* as biocontrol agents against damping-off of forest nurseries. *Plant Dis. Res.* 16: 46-51.

Atilano, R. A., Rich, J. R., Ferris, H. and Menge, J.A. (1976) Effect of *Meloidogyne arenaria* on endomycorrhizal grape (*Vitis vinifera*) rooting. *J. Nematol.* 8 : 278.

Baath, P. and Hayman, D. S. (1983) Plant growth responses to vesicular-arbuscular mycorrhiza XIV. Interaction with Verticillium wilt on tomato plants. *New Phytol.* 95 : 419.

Baker, R., Elad, Y. and Chet, I. (1984) The controlled experiment in the scientific method with special emphasis on biological control. *Phytopathology* 74: 1019-1021.

Bissett, J. (1984) A revision of the genus *Trichoderma.* I Section *Longibrachiatum* sect. nov. *Canadian J. Bot.* 62 : 924-931.

Bissett, J. (1991a). A revision of the genus *Trichoderma* II. Infrageneric classification. *Canadian J. Bot.* 69 : 2357-2372.

Bissett, J. (1991b) A revision of the genus *Trichoderma* III. Section *Pachybasium. Canadian J. Bot.* 69 : 2373-2417.

Bissett, J. (1991c) A revision of the genus *Trichoderma* IV. Additional notes on section *Longibrachiatum. Canadian J. Bot.* 69 : 2418-2420.

Cappellini, R. A. and Peterson, J. N. (1965) Macroconidium formation in submerged cultures by a non-sporulating strain of *Gibberella zeae. Mycologia* 57 : 962-966.

Carling, D. E., Riehle, W. G., Brown, M. F. and Johnson, D. R. (1978) Effects of a vesicular-arbuscular mycorrhizal fungus on nitrate reductase and nitrogenase activities in nodulating and non-nodulating legumes. *Phytopathology* 68 : 1590.

Caron, M. (1989) Potential use of mycorrhizae in control of soilborne diseases. *Canadian J. Plant Pathol.* 11 : 177-179.

Daniel, B. A. and Skipper, H. D. (1982) Methods for the recovery and quantitative estimation of propagules from soil. *In: Methods and Principles of mycorrhizal research* (ed Schenck, N.C.) American Phytopathological Society, St. Paul.

Davis, R. M., Menge, J. A. and Zentmeyer, G. A. (1978) Influence of vesicular-arbuscular mycorrhizae on *Phytopthora* root rot of three crop plants. *Phytopathology* 68 : 1614.

Davis, R. M., Menge, J. A. and Erwin, D. C. (1979) Influence of *Glomus fasiculatum* and soil pathogens on Verticillium wilt of cotton. *Phytopathology* 69 : 453.

Davis, R. M. and Menge, J. A. (1980) Influence of *Glomus fasciculatum* and soil pathogens on Phythophthora root rot of Citrus. *Phytopathology* 70 : 447.

Dehne, H.W. (1982) Interactions between vesicular-arbuscular mycorrhizal fungi and plant pathogens. *Phytopathology* 72 : 1115-1119.

Gerdemann, J. W. and Nicolson, T. H. (1963) Spores of mycorrhizal Endogone species extracted from soil by wet-sieving and decanting, *Tran. British Mycol. Soc.* 46 : 235-244.

Giovannetti, M. and Mosse, B. (1980) An evaluation of techniques to measure vesicular-arbuscular infections in roots. *New Phytòl.* 84 : 489-500.

Harsh, N. S. K. and Ojha, B. M. (1998) A biocontrol formulation for post-emergence damping off of *Moringa pterygosperma*. In: *Proceedings of an International Symposium on microbial Biotechnology for Sustainable development and Productivity*. R. D. University, Jabalpur.

Harsh, N. S. K., Sharma, S., Bhandari, A. S. and Asaiya, A. J. K. (1994) Studies on a new post-emergence damping-off disease of *Moringa pterygosperma*. *J. Trop. For.* 10 : 241-248.

Hooker, J. E., Jaizme-Vega, M. and Atkinson, D. (1994) Biocontrol of plant pathogens using arbuscular mycorrhizal fungi. In: *Impact of Arbuscular Mycorrhizas on Sustainable Agriculture and Ecosystems* (eds S. Gianianazzi and Schuepp) Birkhauser, Swittzerland, pp. 191-200.

Jacobs, D. and Kamoen, O. (1986) Role of cell wall lysing enzymes of *Trichoderma* in antagonism. *Mededelin en Van de Faculteit Landbouwweten schappen Rijksuniversiteit Gemt.* 51 : 751-758.

Jalali, B. L. and Thareja, M. (1985) Plant growth responses to VA mycorrhizal inoculation in soils incorporated with rock phosphate. *Indian Phytopathol.* 38: 306.

Kumar, R. N. (1993) Biological control of damping-off disease in chir pine. *Indian J. Mycol. Plant Pathol.* 23: 237-240.

Manjunath, A. and Bagyaraj, D. J. (1986) Response of black gram chickpea and mungbean to VAM inoculation in immobile nutrients in *Leucaena leucocephala, Plant and Soil* 78 : 445.

Mayer, J. R. (1984) Analyzing the VA mycorhizosphere. In : *Proc. 6th NACOM.* Oregon, USA.

Mc New, G. L. (1960). *Plant Pathology vol. 2*. Academic press, New York.

Phillips, J. M. and Hayman, D. S. (1970) Improved procedure for clearing roots and staining parasitic and vesicular-arbuscular mycorrhizal fungi for rapid assessment of infection. *Trans. British Mycol. Soc.* 55: 158-161.

Rifai, M. A. (1969) A revision of the genus *Trichoderma*. Mycological Papers, C.M.I., U. K.

Schenck, N. C. (1987) Vesicular-arbuscular mycorrhizal fungi and the control of fungal root diseases. In: *Innovative Approaches to Plant Disease Control* (ed I. Chet) John Wiley and Sons, Inc., New York, pp. 179-191.

Schoenbeck, F. (1979) Endomycorrhiza in relation to plant diseases. In: *Soil Borne Plant Pathogens* (eds. Schippers, B. and W. Gams) Academic Press, London.

Arnaud, M. St.., Hmel, C., Caron, M. and Fortin, J. A. (1995) Inhibition of *Pythium ultimum* in roots and growth substrate of mycorrhizal *Tagetes petula* colonized with *Glomus intraradices. Canadian J. Plant Pathol.* 16 : 187-194.

Windham, M. T., Elao, Y. and Baker, R. (1986) A mechanism for increased plant growth induced by *Trichoderma* spp. *Phytopathology* 76 : 518-521.

Young, C. C., Juang, T. C. and Guo, H. Y. (1984) Effects of VA mycorrhizal inoculation on the soybean growth yield, N-fixation and absorption of mineral phosphate. *J. Agr. Assoc. China* 128 : 29.

Zambolim, L. and Schenck, N. C. (1983) Reduction of the effects of pathogenic root-infecting fungi on soybean by the mycorrhizal fungus, *Glomus mosseae. Phytopathology* 73 : 1402-1405.

Zambolim, L. and Schenck, N. C. (1984) Effect of *Macrophomina, Rhizoctonia, Fusarium* and the mycorrhizal fungus *Glomus mosseae* on nodulated and non-nodulated soybeans. *Fitopat. Bras.* 9 : 129-138.

Forest Biotechnology in India © 2008
Editors: S.A. Ansari, C. Narayanan & A.K. Mandal, pp.313-319
Satish Serial Publishing House, Delhi

30

Interaction of AM Fungi and *Azospirillum* Enhances Growth, Survival and Nutrient Uptake in Teak Seedlings

R. K. VERMA AND JAMALUDDIN

Forest Pathology Division, Tropical Forest Research Institute,
PO - RFRC, Jabalpur – 482 021, India

ABSTRACT

Effect of interaction of arbuscular mycorrhizal (AM) fungi and nitrogen fixing bacteria, *Azospirillum* sp. on survival, growth, biomass and P-uptake in teak seedlings was studied in sterilized potting medium. The seedlings were inoculated with AM fungi and *Azospirillum* in the presence of minimal dose of phosphorus. After one year significant differences in survival, growth, biomass and P-uptake were observed among treated and control seedlings. Maximum survival (13.1 times) biomass (198 %) and P-concentration (99%) were observed in combined treatment of AM and *Azospirillum* as compared to uninoculated control, followed by separate AM and *Azospirillum* treatments. These parameters were significantly correlated with root colonization by AM fungi. Based on these observations, teak is classified as highly dependent on arbuscular mycorrhizae for its growth and survival.

Key words: Arbuscular mycorrhizae, dry biomass, endorhizosphere, mycorrhizal dependency, N_2 fixation, root colonization.

INTRODUCTION

Teak (*Tectona grandis* L.) is one of the most important timber yielding tree species of Asian countries particularly India and Myanmar. Teak prefers dry tropical climate and its main growth period is restricted to rainy season. AM fungi colonize teak feeder roots (Verma and Jamaluddin, 1995). Occurrence of various bacteria on the surface or in the cytoplasm of AM spores has been reported (Mosse, 1962; Varma *et al.*, 1981; Ho, 1989; Tilak *et al.*, 1989). Many bacteria are present in the soil and play an important role in growth of plants. In the rhizosphere, the population of these microorganisms increased; and in the mycorrhizal roots (mycorrhizosphere), their population further enhanced (Ames *et al.*, 1984), indicating their role in the growth and development of the plants. The present work is aimed at studying the effect of AM fungi and nitrogen fixing bacteria, *Azospirillum* on survival, growth, biomass and P-uptake in teak seedlings in sterilized potting medium.

MATERIALS AND METHODS

Experimental design

Two factorial experiments were conducted in completely randomized design (CRD) with five replications. Factor one was AM application and factor two was *Azospirillum* application.

Raising of seedlings

The seedlings were raised from seeds collected from Chandrapur, Maharashtra (Clone MHAL A5) in germination trays in sterilized sand. After 45 days, the seedlings were pricked in polythene bags containing 1.672L sterilized substrate, soilrite which contained 0.30% N, 0.08% P, 0.43% K, 15 ppm Zn, 2.74 ppm Cu and 43.19 ppm Mn, on dry weight basis (Sreenivas and Bagyaraj, 1988) and kept on raised iron angle beds. The seedlings were watered with sterile water and fertilized with modified Hoagland's solution with half dose of phosphorus fortnightly (50 ml per seedling).

Preparation of inoculum

An inoculum of AM fungi containing mix species (*Acaulospora scrobiculata, Glomus intraradices, G. etunicatum* and *Scutellospora pellucida*) was multiplied in pot cultures using teak as trap plant. The mycorrhizal roots of trap plants were harvested, washed, sheared to get AM inoculum. *Azospirillum* sp. (strain TF 16) was isolated from teak root collected from local teak plantation. The feeder roots were cut into 1 cm pieces, surface sterilized with 3% H_2O_2, washed with sterile water and suspended in 15 ml culture tubes containing 5 ml semi solid N- free medium containing bromothymol blue and incubated at 30ºC for a week. After 3 days, the bacterium grew into the medium from cut ends of the root segments and characteristic gray-white ring was observed below 2 mm from the surface with change in colour of medium, from green-yellow to blue after one week. The bacterium was separated from the medium by centrifugation (1600xg) and suspended in saline water before use, which contained 10^8 bacteria ml^{-1}.

Inoculation

One ml of sheered root AM inoculum containing 3009 infective propagules (Liu and Luo, 1994) was suspended in 15 ml saline water. 5ml each of AM and bacterial suspension was poured into root zone during pricking of seedlings in polyethylene bags.

Data collection

Data were recorded after one year. The seedlings were uprooted after removing the polythene and cleaning the roots with water carefully. Dry weights were recorded after oven drying plants at 70ºC for 72 h. The percentage of root length colonized was determined after staining roots (Koske and Gemma, 1989) by grid-line-intersect method (Giovanetti and Mosse, 1980).

Detemination of phosphorus in plant material

Phosphorus concentrations in shoots and roots were determined after acid digestion of the samples on a Kjeldhal assembly (Lindner, 1944), by method of Fiske and Subba Row (1925), using a spectrophotometer (Systronic, India; Model 106). The phosphorus concentrations in shoots (P-uptake) was calculated and represented as mg g^{-1} dry weight.

Statistical analysis

The effect of treatments and interactions was analyzed by using two way ANOVA. The means were compared using Duncan's Multiple Range Test for each parameter if "*F*" value was significant at $p<0.05$.

A Pearson Product-Moment Correlation Coefficient among different parameters was worked out.

RESULTS AND DISCUSSION

Application of AM fungi and *Azospirillum* alone or in combination significantly enhanced survival, biomass, root colonization and P-concentration of seedlings after one year (Table 1). Maximum survival was observed in seedlings treated with AM in combination with *Azospirillum* (13.1 times more than the control) followed by AM (10.1 times more) and *Azospirillum* (2.6 times more) (Table 1). Survival of seedlings was highly correlated with root colonization (r=0.94; *p*=0.01).

Table 1: Effect of AM fungi and *Azospirillum* sp. on survival (%), dry weight, root colonization and total phosphorus concentration of teak seedling raised in sterilized soilrite after one year. Values in columns and rows followed by the same letter are not significantly different at *p*=0.05

AM / *Azospirillum* application	**Control**	**AM**	**Mean**
		Survival (%)	
Control	4.5d	45.5b	25.0â
Azospirillum	11.8c	59.1a	35.5á
Mean	8.2y	52.3x	
		Dry weight (g)	
Control	0.59c	1.67a	1.14â
Azospirillum	1.19b	1.76a	1.48á
Mean	0.90y	1.72x	
		Root colonization (%)	
Control	00.0c	72.6b	36.3 â
Azospirillum	00.0c	92.2a	47.6 á
Mean	00.0y	83.9x	
		Total P concentration (mg/g)	
Control	1.56d	2.46b	2.01 â
Azospirillum	1.91c	3.11a	2.51 á
Mean	1.74y	2.79x	

Both AM and *Azospirillum* application significantly enhanced dry weights of seedlings. *Azospirillum* treatment alone increased 102% dry biomass, and 198% when applied in combination with AM fungi compared to the control. Maximum root colonization was obtained in AM along with *Azospirillum* treatment which was 27.7% more than AM alone treatment. There was no root colonization in uninoculated control seedlings (Table 1).

Phosphorus concentration in plant tissue was maximum in seedlings which received AM along with *Azospirillum* (99% more than the control) followed by AM (58% more) and *Azosirillum* alone treatments (22% more) (Table1). Root colonization was highly correlated with P concentration (r=0.91 at p=0.01).

The AM symbiosis is known to promote acquisition of mineral nutrients especially P by host plants (Bolan, 1991; Copper and Tinker 1978; Lambert *et al.* 1979), as a result of additional growth of AM hyphae that enlarges root-soil interface, facilitating efficient exploration of soil around the root (Jakobsen *et al.*, 1992; Li *et al,*. 1991). On P-uptake in AM applied seedlings, the present study reaffirmed that improved growth in AM treated plants was mediated through improved uptake of P and other nutrients. P-uptake in teak seedlings was significantly correlated with stem height and dry biomass of shoots. Higher root colonization facilitated improved P-uptake that resulted in enhanced height and dry biomass production in the plants having higher root colonization. Development of mycorhizae in feeder roots and spread of hyphae of AM fungi in the rhizosphere may help in tapping of additional nutrients from soil by mycorrhizal plants as compared to uninoculated control. More accumulation of P in plant dry biomass was due to enhanced uptake of P from soil with the help of AM fungi. Enhancement in P-uptake in AM-treated seedlings may possibly also be due to mineralization of organic P by AM fungi (Jayachandran *et* al., 1992) and by *Azospirillum* (Seshadri *et al.*, 2000).

Soilrite, the potting mix used in this experiment had very poor nutrient content (Sreenivas and Bagyaraj, 1988). The experimental seedlings also received low doses of nutrient, especially P, in the form of nutrient solution and showed a very poor growth. It also appears that soilrite is also not a good substrate for raising teak seedlings as the percentage of survival was very low, 4.5% in the control (Table 1).

Enhancement in dry biomass due to *Azospirillum* treatment may be due to bacterial N_2 fixation, which was utilized by the teak seedlings. There is synergistic interaction in AM fungi and *Azospirillum* as both enhanced dry biomass and root colonization significantly. Many plant growth promoting rhizobacteria are also reported as mycorhization helper bacteria. Since this bacterium is also increasing mycorhization in teak, which is correlated with dry weight (r=0.81 at p=0.01), it may also be considered as mycorhization helper bacteria. Based on above observations, teak is classified as highly dependent species on arbuscular mycorrhizae and *Azospirillum* for its growth and survival.

ACKNOWLEDGEMENTS

We wish to thank to Dr. A. K. Mandal, Director, Tropical Forest Research Institute, Jabalpur for providing the facilities and Mr. Ashok K. Thakur for his assistance during the course of the present study. Thanks are also due to Dr. Ching Y. Li of USDA for demonstration of method to isolate *Azospirillum* from endorhizosphere of root during his visit to our laboratory.

References

Ames, R. N., Reid, C. P. P. and Ingham, E.R. (1984) Rhizosphere bacterial population response to root colonization by a vesicular-arbuscular mycorrhizal fungus. *New Phytol.* 96 : 555-563.

Bolan, N. S. (1991) A critical review on the role of mycorrhizal fungi in the uptake of phosphorus by plants. *Plant Soil* 134 : 189-207.

Copper, K. M. and Tinker, P. B. (1978) Translocation of and transfer of nutrients in vesicular-arbuscular mycorrhizas. II. Uptake and translocation of phosphorus zinc and sulphur. *New Phytol.* 81 : 43-51.

Fisk, C.H. and Subba Row, Y. (1925) The calorimetric determination of phosphorus. *J. Biol. Chem.* 66 : 375-400.

Giovanetti, M. and Mosse, B. (1980) An evaluation of techniques for measuring vesicular arbuscular infection in roots. *New Phytol.* 84 : 489-500.

Ho, I. (1988) Interaction between VA-mycorrhizal fungus and *Azotobacter* and their combined effect on growth of tall fescue. *Plant Soil* 105 : 291-293.

Jakobsen, I., Abbott, L. K. and Robson, A. D. (1992) External hyphae of vesicular-arbuscular mycorrhizal fungi associated with *Trifolium subterraneum* L. I. Spread of hyphae and phosphorus inflow into roots. *New Phytol.* 120 : 371-380.

Jayachandran, K., Schwab, A. P. and Hetrick, B. A. B. (1992) Mineralization of organic phosphorus by vesicular-arbuscular mycorrhizal fungi. *Soil Biol. Biochem.* 24: 897-903.

Koske, R. E. and Gemma, J. N. (1989) A modified procedure for staining roots to detect VA-mycorrhiza. *Mycol. Res.* 92 : 496-505

Lambert, D. H., Baker, D. E., and Cole, H. Jr. (1979) The role of mycorrhizae in the interaction of phosphorus with zinc, copper and other elements. *Soil Sci.* 43 : 976-980.

Li, X. L., George, E. and Marschner, H. (1991) Extension of the phosphorus depletion zone in VA mycorrhizal white clover in a calcareous soil. *Plant Soil* 136: 41-48.

Lindner, R. C. (1944) Rapid analytical methods for some of the more common inorganic constituents of plant tissue. *Plant Physiol.* 19 : 76-89.

Liu, R. J. and Luo, X. S. (1994) A new method to quantify the inoculum potential of arbuscular mycorrhizal fungi. *New Phytol.* 128 : 89-92.

Mosse, B. (1962) The establishment of vesicular arbuscular mycorrhiza under aseptic conditions. *J. Gen. Microbiol.* 27 : 509-520.

Seshadri, S., Muthukumarasamy, R., Lakshminarsimhan, C. and Ignasimuthu, S. (2000) Solubilization of inorganic phosphate by *Azospirillum halopraeferans. Curr. Sci.* 79 : 565-567.

Sreenivas, M. N. and Bagyaraj, D. J. (1988) Selection of a suitable substrate for mass multiplication of *Glomus mosseae. Plant Soil* 109 : 125-127.

Tilak, K. V. B. R., Li, C. Y. and Ho, I. (1989) Occurrence of nitrogen fixing *Azospirillum* in vesicular-arbuscular mycorrhizal fungi. *Plant Soil* 116 : 286-288.

Varma, A. K., Singh, K. and Lall, V. K. (1981) Lumen bacteria from endomycorrhizal spores. *Curr . Microbiol .* 6 : 207-211.

Verma, R. K. and Jamaluddin (1995) Association and activity of arbuscular mycorrhizae of *teak (Tectona grandis)* in central India. *Indian For.* 121 : 533-539.

□□□